BASIC CONCEPTS
IN
ENVIRONMENTAL
MANAGEMENT

BASIC
CONCEPTS
── IN ──
ENVIRONMENTAL
MANAGEMENT

KENNETH M. MACKENTHUN

CRC Press
Taylor & Francis Group
Boca Raton London New York

CRC Press is an imprint of the
Taylor & Francis Group, an **informa** business

CRC Press
Taylor & Francis Group
6000 Broken Sound Parkway NW, Suite 300
Boca Raton, FL 33487-2742

First issued in paperback 2019

ISBN-13: 978-1-56670-380-2 (hbk)
ISBN-13: 978-0-367-39998-6 (pbk)

Library of Congress Cataloging-in-Publication Data

Mackenthun, Kenneth Marsh, 1919–
    Basic concepts in environmental management / Kenneth M.
Mackenthun.
        p.   cm.
    Includes index.
    ISBN 1-56670-380-8 (alk. paper)
       1. Environmental management--United States. 2. Environmental
policy--United States. I. Title.
GE310.M33   1999
363.7'056'0973--dc21

                                                98-37469
                                                  CIP

Library of Congress Card Number 98-37469

**Visit the Taylor & Francis Web site at
http://www.taylorandfrancis.com**

**and the CRC Press Web site at
http://www.crcpress.com**

Dedicated to those who support, foster, and promote the development of an environmental ethic in the general population.

Author's Profile

Kenneth M. Mackenthun, in a 52-year career in water pollution investigations and control, water resources management, hazardous waste management, environmental assessment, and regulation development, has produced over 100 technical publications, including 11 books. He graduated cum laude with a degree in Biology from Emporia (Kansas) College and obtained an M.A. in Zoology from the University of Illinois in 1946.

Mr. Mackenthun worked for the Wisconsin Department of Conservation and Wisconsin State Board of Health for 16 years where he served as Chief Biologist. He served 20 years with the U.S. Environmental Protection Agency (EPA) and its predecessor, the U.S. Public Health Service. He was Director of the Criteria and Standards Division (Water) and Acting Deputy Administrator for Water Planning and Standards where he supervised 220 employees in 4 Divisions. He was responsible for the Clean Water Act Estuarine Report to Congress; In-Place Pollutant Removal; National Water Quality Standards; Water Quality Criteria Development; report on Methods and Procedures to Restore Lakes; Maintaining the Toxic Pollutants List; Hazardous Substances Discharges; Vessel Sewage Discharge Regulations; Aquaculture Program and Regulations; Clean Lakes Restoration Program; and Dredge and Fill Materials Discharge Program.

He served as Adjunct Professor at The American University, Washington, D.C., teaching several courses in the off-campus graduate degree program. He operated his own consulting firm for several years, and served a number of environmental consulting firms, the most recent being ADI Technology Corporation in Crystal City, VA, where he provided environmental consultation to the U.S. Navy's submarine program. He is a recognized expert in environmental assessments and water regulations.

His experience includes 16 years in state government service, 19 years in federal government service, and 17 years serving various environmental consulting firms.

The contents of this book do not supersede or replace the regulatory requirements of federal, state, or local authorities.

Table of Contents

List of Tables xiv

Chapter 1
Introduction 1
1.1 Purpose 1
1.2 Contents 1
1.3 Background 2
1.4 Environmental Ethic 4

Chapter 2
Fundamentals 5
2.1 Acronyms 5
    2.1.1 Commonly Used Acronyms 5
    2.1.2 Audits 6
    2.1.3 Citizen's Suit 6
2.2 U.S. Code 6
2.3 Code of Federal Regulations 6
    2.3.1 Codification of Federal Regulation 6
    2.3.2 Congressional Record 7
    2.3.3 Effluent Guidelines 7
    2.3.4 Enforcement 8
    2.3.5 Environmental Investigation 8
    2.3.6 Executive Orders 9
    2.3.7 Federal Register 9
    2.3.8 Freedom of Information 9
    2.3.9 Legislative History 9
2.4 Public Laws 9
    2.4.1 Permits 9
2.5 Regulations 11
    2.5.1 Risk Assessment 14

Chapter 3
Assessments and Impacts 15
3.1 The National Environmental Policy Act of 1969 15
3.2 Farmland Protection Policy Act of 1991 17
3.3 Environmental Effects Abroad of Major Federal Actions 18

Chapter 4
Water 19
4.1 Surface Fresh Water 20
4.2 Non-Indigenous Aquatic Nuisance Prevention and Control Act of 1990 28
4.3 National Invasive Species Act of 1996 28

4.4   Ground Water 29
       4.4.1 Safe Drinking Water Act of 1974 29
4.5   Marine Water 31
       4.5.1 Marine Protection, Research, and Sanctuaries Act
              of 1972 (MPRSA) 31
4.6   Convention for the Prevention of Pollution from Ships (MARPOL) 32
       4.6.1 Annex 1—Prevention of Pollution by Oil 33
       4.6.2 Annex 2—Prevention of Bulk Hazmat Pollution 33
       4.6.3 Annex 3—Prevention of Containerized Hazmat Pollution 33
       4.6.4 Annex 4—Prevention of Pollution by Sewage 33
       4.6.5 Annex 5—Prevention of Pollution by Garbage 34
4.7   Act to Prevent Pollution from Ships (APPS) of 1980 34

Chapter 5
Air 37
5.1   Clean Air Act of 1970 (CAA) 37
5.2   Asbestos 41
5.3   Other Sections of the Clean Air Act 43
5.4   Air Glossary 43

Chapter 6
Land 49
6.1   Resource Conservation and Recovery Act (RCRA) 49
6.2   The Comprehensive Environmental Response, Compensation,
       and Liability Act of 1980 (CERCLA, Superfund) 53
       6.2.1 A Simplified Concept 53
       6.2.2 Principal Provisions 54

Chapter 7
Toxic and Hazardous Materials and Wastes 59
7.1   Toxic Substances Control Act of 1976 (TSCA) 59

Chapter 8
Other Environmental Laws 61
8.1   Executive Order 12856 61
8.2   EPCRA, Emergency Planning and
       Community Right-To-Know Act of 1986 61
8.3   Pollution Prevention Act of 1990 62
8.4   Endangered Species Act 64
8.5   Marine Mammal Protection Act of 1972 65
8.6   Federal Insecticide, Fungicide, and Rodenticide Act (FIFRA) 68
8.7   Fish and Wildlife Coordination Act 70
8.8   Migratory Bird Treaty Act 71
8.9   Bald Eagle Protection Act 71
8.10   National Historic Preservation Act 71
8.11   Coastal Zone Management Act of 1972 71

Chapter 9
Compliance 73

Chapter 10
Ecosystem Recovery 77

Chapter 11
Pollution Prevention 81

Chapter 12
Review Questions and Answers 83
12.1   Questions 83
12.2   Answers 88

Chapter 13
Safety and Precautions 95

Chapter 14
Affected Environments 99
14.1   Temperature 99
14.2   Dissolved Oxygen 102
14.3   pH 103
14.4   Light 103
14.5   Flow 103
14.6   Major Nutrients 104
14.7   Minor Nutrients 105
14.8   Pollutants 105
            14.8.1 Inorganic Silts 106
            14.8.2 Toxic Materials 106
            14.8.3 Organic Pollution 106
14.9   Water Quality Investigation 108
            14.9.1 Objectives 108
            14.9.2 Planning 109
            14.9.3 Data Collection 112
            14.9.4 Lakes 113
            14.9.5 Streams 115

Chapter 15
The Environmental Career 117
15.1   Keys to Career Success 117
15.2   Diligent Work Habits 117
15.3   Communication Ability 117
15.4   Investigative Reporting 118
            15.4.1 Outline 118
            15.4.2 Organization 118
            15.4.3 Report Development 119
            15.4.4 Review and Final Report 120

15.5 Knowledge of Job-Related Subject Matter 121
15.6 Personal Drive to Present and Publish Information 121
15.7 Career Opportunities ......·............................... 122
     15.7.1 Government Service 122
     15.7.2 Industry 123
     15.7.3 Consulting Activities 123
15.8 Handy Trouble Numbers 124

References Cited and Selected Reading 125

Index 127

Tables

2.1 List of Commonly Used Acronyms 5
2.2 The Titles of the U.S. Code 7
2.3 The Titles of the Code of Federal Regulations 8
2.4 Selected Titles of Environmental Laws 10
2.5 Selected Significant Environmental Regulations 12
3.1 Format for an Environmental Impact Statement 16
3.2 EIS Information Related to Effects on the Affected Environment 17
4.1 The 65 Toxic Water Pollutants 23
4.2 Sensitive Ecological and Structural Areas 25
4.3 Process for Developing a Regulation, Effluent Guideline,
    and NPDES Permit 26
4.4 Diagram Summarizing the CWA Regulatory Process 27
5.1 Similarities Between the Clean Air Act and Clean Water Act
    in Methods to Control Pollution 37
5.2 Hazardous Air Pollutants 40
6.1 Toxicity Characteristics Leaching Procedure Maximum
    Concentration of Contaminants 50
9.1 Sample Questions for a Hazardous Waste Management
    Self-Audit 73

1 Introduction

Russell Train, first Chairman of the Council on Environmental Quality and second Administrator of the U.S. Environmental Protection Agency (EPA) is quoted as saying, "Whether we like it or not, continued economic and population growth guarantee that environmental issues are going to become more urgent and complex, not less."

1.1 PURPOSE

This book contains the information you need to know to begin, sustain, and extend a career in environmental and natural resource protection in federal, state, and local governments; industry; and consulting activities. The need to have knowledgeable career people in the environmental field is demonstrated by the heightened public interest in environmental matters, the increasing regulatory compliance requirements, regulatory violations that could and should be eliminated, and the vast number of federal and state environmental regulations promulgated since the EPA was formed in 1970.

The environmental advantage is in maintaining a knowledge of, and being conversant with, environmental regulations and laws, the history of how the laws came into existence, compliance requirements, and the environments to be protected. The personal advantage is considerable and is measurable in career enhancement and personal monetary compensation.

The author's career and interests were first focused on pollution and the water environment in the 1940s and 1950s prior to a general public awareness of the need to control environmental pollution. It extended through the decade of the 1960s (the era of field investigations and demonstrations of the nature and extent of pollution), the decade of the 1970s (the era of federal pollution control laws and regulatory development), and into the 1990s (the era of regulatory refinement and the treatment of non-point source pollution problems previously left unresolved).

1.2 CONTENTS

Let us move ahead to the organization of the text in this book. The next chapter addresses fundamental and environmental concepts that are necessary to an understanding of information related to environmental laws and regulations. Chapter 3 addresses environmental assessment and impacts, National Environmental Policy Act (NEPA) requirements, and the NEPA process, which is essential to sound environmental planning and management.

The next three chapters discuss the air, land, and water resources and introduce the reader to 11 associated federal laws and their respective requirements. There are regulations that are associated with and implement various sections of the laws, and these are not neglected.

1

Chapters 7 and 8 deal with multimedia toxic and hazardous wastes and laws and regulations that do not fit well with the previous discussions. Chapter 9 discusses the all important matter of environmental compliance. Chapter 10 reminds us that ecosystems will recover when pollution is curtailed. Chapter 11 addresses the current important words, pollution prevention. Chapter 12 provides review questions and answers related to the principal environmental laws and regulations contained in the previous chapters. Chapter 13 discusses safety precautions and features that are necessary when working with toxic and hazardous materials and wastes. Chapter 14 provides a discussion of the vulnerability of the environments most often impacted by pollution occurrences. Chapter 15 records experiences and discussion of experiences associated with a career in government service, industry, and consulting.

1.3   BACKGROUND

Generally, our environment is much cleaner now than it was in the 1950s and 1960s. The decade of the 1960s was the decade that awakened the people of the United States to the fact that air and water pollution must be better controlled. Many lakes, including one of the Great Lakes, were declared "dead" because they were so enriched with nutrients such as nitrogen and phosphorus that massive algal growths developed and later decomposed liberating hydrogen sulfide that permeated the air and blackened white lead paint on shoreside houses that were three or four blocks from the lake. Opened-windowed sleeping was not practical during the summer nights. Sport fishing was no longer an enjoyable sport. At least one river near Chicago contained sufficient floating oil that it burst into flames when a cigarette was tossed into it. Another river in the northeast had floated so many logs from the timber area to a paper mill that its bottom was covered with decomposing wood chips. A chemical, sodium nitrate, was literally shoveled into the river to add dissolved oxygen from the nitrification process. The environment is better now, but there still is pollution and vigilance must still be maintained to ensure that backsliding will not occur and our pollution control efforts continue with the momentum they have achieved.

States are required by Section 305 (b) of the Clean Water Act to submit a biennial report to the EPA that describes the water quality of the state. EPA, in turn, prepares a National Water Quality Inventory, which is submitted to the Congress biennially. The latest EPA report (EPA, 1995) states that the United States has 3.5 million miles of rivers, and streams; 40.8 million acres of lakes, ponds, and reservoirs; 34,388 square miles of estuaries; 58,000 miles of ocean shoreline, 5,559 miles of Great Lakes shoreline, and 277 million acres of wetlands such as marshes, swamps, bogs, and fens. Alaska boasts 170 million acres of this wetland total.

According to the EPA report, rivers are polluted by bacteria, siltation, nutrients, oxygen-depleting substances, metals, habitat alterations, and suspended solids. The leading source of pollution is from agriculture with municipal point sources being second. Lakes are polluted by nutrients, siltation, oxygen-depleting substances, metals, suspended solids, pesticides, and priority organic toxic pollutants. Agriculture, again, was the leading source of pollution with municipal point sources in second place.

The leading pollutant for the Great Lakes is toxic organic chemicals, primarily polychlorinated biphenyls (PCBs), followed in much smaller measure by pesticides and non-priority organic chemicals. The principal leading sources of pollution are air pollution, discontinued discharges, and contaminated sediments.

Estuaries are polluted principally by nutrients, bacteria, and oxygen-depleting substances, which are provided by urban runoff and storm sewers, municipal point sources, agriculture and industrial point sources, in that order. Wetlands integrity is affected by sediments, flow alterations, habitat alterations, and filling and draining. Agriculture, urban runoff, and hydrologic modification are the principal sources. Ground water is affected by leaking underground storage tanks (an estimated 139,000 tanks have leaked), agricultural activities (where 1.1 billion lb of pesticides are applied to agricultural lands annually), superfund sites, and septic tanks. The most common ground water contaminants are petroleum compounds, nitrates, metals, volatile organic compounds, and pesticides.

Pursuant to Section 201 of the National Environmental Policy Act, the President shall transmit to the Congress an Environmental Quality Report. The current report for 1994 and 1995 is the 25th Anniversary Report by the Council on Environmental Quality. This report indicates that carbon monoxide emissions from 1970 to 1994 declined from 128 to 98 million tons per year or 23 percent. Lead emissions are down 98 percent over the 1970 to 1994 period and 75 percent over the 1985 to 1994 period. Nitrogen oxide emissions are up 14 percent from 10.6 to 23.6 million tons per year during the 1970 to 1994 period. High levels of ozone persist in many heavily populated areas, including much of the Northeast, the Texas Gulf Coast, and Los Angeles. It is estimated that about 50 million people lived in counties with ozone levels above the national standard in 1994. Over the 1970 to 1994 period, emissions of sulfur dioxide were down 32 percent.

The conversion of wetlands to other uses has slowed over the past several decades, dropping from an average of 690,000 acres per year in the 1954 to 1974 period to about 423,000 acres annually in the 1974 to 1983 period. The total wetland acreage in the United States in the mid-1980s was about 103.24 million acres. This is considerably less than the EPA report discussed earlier.

Municipal solid waste generation has grown steadily and is expected to grow even more. From 1960 to 1994, waste generation increased from 88 million tons to 209 million tons per year. Per capita generation rose from 2.7 lb/d in 1960 to 4.4 lb/d in 1994. It is projected to hold steady at 4.4 lb per capita per day through the year 2000, but increase to 4.7 lb per capita per day by the year 2010.

The 25th Anniversary Report states that the number of lakes, rivers, and other U.S. waterways where consumers have been advised to avoid or limit consumption of trout, salmon, or other fish species because of chemical contamination rose from 1,278 in 1993 to 1,740 in 1995. The contaminated fish advisories were issued for fish flesh concentrations of mercury, PCBs, chlordane, dioxins, and DDT. The increased number of contaminated fish advisories was owing to mercury and PCB concentrations in fish flesh. The fish advisories, issued for more than 1,700 water bodies, represent a 14 percent increase over the previous year.

1.4   ENVIRONMENTAL ETHIC

There is evidence that an environmental ethic is gaining ground among the general population. People are feeling better about their environment. Leopold (1949) attempted to establish an ecological conscience among his contemporaries. Leopold's message has persisted. That message is needed more today than ever before to counteract any effort at backsliding. Hopefully, a couple of decades from now, the environmental ethic will be even stronger among the general population. The human population will be more at peace with the requirements of its environment. The environmental ethic will dominate that relationship.

Regulations are required to keep societal goals focused on future attainment. In 1992, Mackenthun and Bregman published certain principles related to environmental regulations. It is worthwhile to repeat them here:

* The purpose of an environmental regulation is to adequately protect, in a broad sense, the environment for man.
* Because of the infinite variation in biological and chemical reactions to environmental change and perturbations, regulations must allow flexibility for adjustment to a particular circumstance. The burden of proof for a more restrictive regulation logically rests with government; the burden of proof for justifying regulatory relaxation, where the need is demonstrated through investigation and assessment, rests with the discharger.
* Political and legislative deadlines often have fostered benign cynicism because of their unrealistic nature in a regulatory world of abundant checks and balances.
* Determining the current state-of-the-art knowledge related to a particular subject is a time-consuming endeavor. The search for an answer in the realm of the unknown is research operating though undeterminable time.

2 Fundamentals

There are certain fundamental concepts, including a glossary of regulatory terms, that contribute to a more complete understanding of the laws, regulations, and compliance issues associated with environmental and natural resource protection. These are presented in this chapter.

2.1 ACRONYMS

2.1.1 COMMONLY USED ACRONYMS

TABLE 2.1
List of Commonly Used Acronyms

APPS	Act to Prevent Pollution from Ships	NAMS	National Air Monitoring Stations
BMP	Best Management Practice	NCP	National Contingency Plan
CAA	Clean Air Act	NEPA	National Environmental Policy Act
CATEX	Categorical Exclusion (NEPA)	NESHAP	National Emission Standard for
CERCLA	Comprehensive Environmental		Hazardous Air Pollutants
	Response, Compensation, and	NO_x	Nitogen Oxides
	Liability Act (Superfund)	NPL	National Priority List
CFC	Chlorofluorocarbon	OPA	Oil Pollution Act
CFR	Code of Federal Regulations	O_3	Ozone
CO	Carbon Monoxide	OSHA	Occupational Safety and Health Act
CWA	Clean Water Act	PEL	Permissible Exposure Level
CZMA	Coastal Zone Management Act	Pb	Lead
EA	Environmental Assessment (NEPA)	PCB	Polychlorinated Biphenyls
EO	Executive Order	PPA	Pollution Prevention Act
EIS	Environmental Impact Statement	RCRA	Resource Conservation and
	(NEPA)		Recovery Act
EPCRA	Emergency Planning and	SIP	State Implementation Plan
	Community	SDWA	Safe Drinking Water Act
	Right-to-Know Act	SWDA	Solid Waste Disposal Act
HAZMAT	Hazardous Material	SLAMS	State and Local Air Monitoring
HW	Hazardous Waste		Stations
MPRSA	Marine Protection, Research,and	TSCA	Toxic Substances Control Act
	Sanctuaries Act	SO_x	Sulfur Oxides
NAA	Nonattainment Area	WQC	Water Quality Criteria
NAAQS	National Ambient Air Quality	WQS	Water Quality Standards
	Standards		

2.1.2 AUDITS

An inspection by one or more persons to compare program operation with plans, standard procedures, or regulations. It may be conducted by persons within an organization, outside an organization, or by an official governmental unit. An audit may be conducted on quality assurance procedures as followed by an analytical laboratory, project program management, or any functional entity of a program. Frequent self-audits generally maintain a level of environmental compliance satisfactory to avoid receiving a notice of violation from a regulatory agency. An audit is tantamount to an environmental report card.

2.1.3 CITIZEN'S SUIT

A civil action commenced by any citizen against any person who is alleged to be in violation of a permit, or against the Administrator of EPA where there is alleged failure to perform any act or duty pursuant to applicable law, which is not discretionary. Citizen's suits are permitted under many environmental laws, such as Section 505 of the Clean Water Act.

2.2 U.S. CODE

The U.S. Code is a consolidation and codification of all laws of the United States. It is prepared under the supervision of the Law Revision Counsel of the House of Representatives. The Code is *prima facie* law and constitutes a library of approximately 300 volumes. This classification system of U.S. law was begun in 1926. The titles have not changed significantly since. See Table 2.2.

Title 15 contains the National Historic Preservation Act and the Toxic Substances Control Act. Title 16 contains the Archaeological and Historic Preservation Act, Coastal Zone Management Act, Endangered Species Act, Fish and Wildlife Coordination Act, and the Marine Mammal Protection Act. Title 29 contains the Occupational Safety and Health Act. Title 33 contains the Act to Prevent Pollution from Ships; Clean Water Act; Marine Protection, Research and Sanctuaries Act; Oil Pollution Act; and River and Harbor Act. Title 42 contains the Clean Air Act; Comprehensive Environmental Response, Compensation and Liability Act; Emergency Planning and Community Right-To-Know Act; National Environmental Policy Act; Noise Control Act; Pollution Prevention Act; Radon Gas and Indoor Air Quality Research Act; Resource Conservation and Recovery Act; and the Safe Drinking Water Act. Title 49 contains the Hazardous Materials Transportation Act.

2.3 CODE OF FEDERAL REGULATIONS

2.3.1 CODIFICATION OF FEDERAL REGULATION

The final rules promulgated by federal agencies in the Federal Register are codified annually in the Code of Federal Regulations (CFR). In codification, only the rule is retained; the preamble to the rule, with its discussion and explanation, is not codified. Environmental regulations are codified in 40 CFR as of July 1 each year. There are approximately 225 volumes of CFRs for all federal regulations, and 18 CFR volumes for the environmental regulations.

TABLE 2.2
The Titles of the U.S. Code

1. General Provisions	15. Commerce and Trade	Procedure	39. Postal Service
2. The Congress	16. Conservation	29. Labor	40. Public Buildings, Property, and
3. The President	17. Copyrights	30. Mineral Lands and Mining	Works
4. Flag and Seal, Seat of Government, and the States	18. Crimes and Criminal Procedure	31. Money and Finance	41. Public Contracts
	19. Customs Duties	32. National Guard	42. The Public Health and Welfare
5. Government Organization and Employees	20. Education	33. Navigation and Navigable Waters	43. Public Lands
	21. Food and Drugs		44. Public Printing and Documents
6. Surety Bonds	22. Foreign Relations and Intercourse	34. Navy (See Title 10, Armed Forces)	45. Railroads
7. Agriculture	23. Highways		46. Shipping
8. Aliens and Nationality	24. Hospitals and Asylums	35. Patents	47. Telegraphs, Telephones, and Radiotelegraphs
9. Arbitration	25. Indians	36. Patriotic Societies and Observances	
10. Armed Forces	26. Internal Revenue Code		48. Territories and Insular Possessions
11. Bankruptcy		37. Pay and Allowances of the Uniformed Services	
12. Banks and Banking	27. Intoxicating Liquors		49. Transportation
13. Census	28. Judiciary and Judicial		50. War and National Defense
14. Coast Guard		38. Veterans' Benefits	

The *Federal Register* (FR) began publication in 1934. The federal regulations were codified in the Code of Federal Regulations in 1937. The CFR titles initially were the same as those for the U.S. Code. Some have been legislated out of existence, e.g., numbers 2 and 6 are now reserved; new titles have been added, e.g., number 40; and some have been changed because of agency preference. For both the U.S. Code and the CFR, 26 titles remain the same. See Table 2.3.

2.3.2 CONGRESSIONAL RECORD

The Congressional Record is a daily weekday publication containing all happenings in the Congress for the date of publication.

2.3.3 EFFLUENT GUIDELINES

Effluent guidelines provide National Standards of Performance for industrial discharges and pretreatment standards to industries that discharge to publicly-owned treatment works. Effluent limitations may be expressed in a concentration or in a weight per weight of product produced. When available for a Standard Industrial Classification (SIC) code, they are used in developing National Pollutant Discharge Elimination System (NPDES) permits.

TABLE 2.3
The Titles of the Code of Federal Regulations

1. General Provisions	16. Commercial Practices	28. Judicial Administration	41. Public Contracts and Property Management
2. Reserved	17. Commodity and Securities Exchanges	29. Labor	42. Public Health
3. The President		30. Mineral Resources	43. Public Lands: Interior
4. Accounts	18. Conservation of Power and Water Resources	31. Money and Finance: Treasury	44. Emergency Management and Assistance
5. Administrative Personnel		32. National Defense	
6. Reserved	19. Customs Duties	33. Navigation and Navigable Waters	45. Public Welfare
7. Agriculture	20. Employee's Benefits	34. Education	46. Shipping
8. Aliens and Nationality	21. Food and Drugs	35. Panama Canal	47. Telecommunication
9. Animals and Animal Products	22. Foreign Relations	36. Parks, Forests, and Public Property	48. Federal Acquisition Regulations System
10. Energy	23. Highways	37. Patents, Trademarks, and Copyrights	
11. Federal Elections	24. Housing and Urban Development		49. Transportation
12. Banks and Banking		38. Pensions, Bonuses, and Veterans' Relief	50. Wildlife and Fisheries
13. Business Credit and Assistance	25. Indians	39. Postal Service	
14. Aeronautics and Space	26. Internal Revenue	40. Protection of Environment	
15. Commerce and Foreign Trade	27. Alcohol, Tobacco Products, and Firearms		

2.3.4   ENFORCEMENT

Enforcement is the ultimate action to seek compliance with law or regulations and to assess damages for past non-compliance. It may involve compliance inspections or monitoring, sampling and analyses, gathering of other evidence, administrative actions, civil actions, or criminal actions. Administrative actions may be either informal or formal. Informal administrative actions are notices of non-compliance or warning letters. Formal administrative actions are legal actions that result in an order requiring the violating party to correct the violation, and in most cases, to pay a civil penalty that is commensurate with the seriousness of the violation. Civil actions are taken in the U.S. court. Criminal actions are taken when a person has knowingly and willfully committed a violation of the law.

2.3.5   ENVIRONMENTAL INVESTIGATIONS

Environmental investigations are studies or surveys to assess the effects on the receiving media of pollution or environmental perturbations. Generally, they involve advanced planning, sampling location determination, sampling, sample analyses, data interpretation, and report preparation.

2.3.6 EXECUTIVE ORDERS

Executive orders are orders signed by the President directing one or more federal agencies to accomplish certain actions. An example is EO 12856 that directed all applicable federal agencies to comply with the requirements of the Emergency Planning and Community Right-To-Know and Pollution Prevention acts. Prior to this EO, federal agencies were excluded from such compliance because of the Standard Industrial Code (SIC) applicability.

2.3.7 FEDERAL REGISTER

The *Federal Register* is the regulatory newspaper of the federal government. It is issued daily on weekdays, excluding holidays. It contains Presidential documents, including Executive Orders; Proposed Rules; Interim Final and Final Rules; regulatory policy decisions; and notices of meetings and other regulatory events, including the availability of applicable technical and guidance documents. Each agency of the federal government is represented. The *Federal Register* is properly cited as: volume number, FR, the first page number of the citation, and this may be followed by a comma and the date of publication (e.g., 63 FR 12418, March 13, 1998). The FR is published by the Office of Federal Register, National Archives and Records Administration.

2.3.8 FREEDOM OF INFORMATION

The freedom of Information Act provides that government information other than privileged or confidential information shall be made available, upon written request, to the requestor. A reasonable fee may be charged for document duplication.

2.3.9 LEGISLATIVE HISTORY

A document specific to a public law that records statements made by members of the Congress, including deliberations and changes made by the Conference Committee, related to various sections of the respective Senate and House Bills as they are passed and molded into a Bill for the consideration of the President.

2.4 PUBLIC LAWS

An example is Pub. L. 92-500, the Federal Water Pollution Control Act Amendments of 1972, which was the 500th law of the 92 Congress. There are approximately 80 public laws that are associated with environmental protection. See Table 2.4.

2.4.1 PERMITS

Permits are documentary authorization, with compliant conditions, for a legal discharge to the environment. Permits may be general (covering all activities of a similar nature in a geographical area) or specifically focused on one point source discharge or emission. Upon application for a permit, the applicant may be required

TABLE 2.4
Selected Titles of Environmental Laws

Abandoned Shipwreck Act of 1987, 43 USC 2101
Acid Precipitation Act of 1980, 42 USC 8901
Act to Prevent Pollution from Ships of 1980, 33 USC 1901
Agricultural Act of 1970, 42 USC 3122
American Indian Religious Freedom Act of 1978, 42 USC 1996
Anadromous Fish Conservation Act of 1965, 16 USC 757a
Antiquities Act of 1906, 16 USC 431
Archaeological and Historic Preservation Act of 1980, 16 USC 469
Archaeological Resources Protection Act of 1979, 16 USC 470a
Arctic Research and Policy Act of 1984, 15 USC 4101
Aviation Safety and Noise Abatement Act of 1979, 49 USC 2101
Bald Eagle Protection Act, 16 USC 668
Clean Air Act of 1970, 42 USC 7401
Clean Water Act of 1977, 33 USC 1251
Coastal Barrier Resources Act of 1982, 16 USC 3501
Coastal Zone Management Act of 1972, 16 USC 1451
Comprehensive Environmental Response, Compensation and Liability Act of 1980, 42 USC 9601
Conservation Programs on Military Installations (Sikes Act), 16 USC 670a
Deepwater Port Act of 1974, 33 USC 1501
DOD Environmental Restoration Program, 10 USC 2701
Earthquake Hazards Reduction Act of 1977, 42 USC 7701
Emergency Energy Conservation Act of 1979, 42 USC 6261, 6422, 8501
Emergency Planning and Community Right-to-Know Act of 1986, 42 USC 11001
Endangered Species Act of 1973, 16 USC 1531
Environmental Quality Improvement Act of 1970, 42 USC 4371
Estuary Protection Act of 1968, 16 USC 1221
Farmland Protection Policy Act of 1981, 7 USC 4201
Federal Facility Compliance Act of 1992, 42 USC 6901 note, 6908
Federal Insecticide, Fungicide and Rodenticide Act of 1947, 7 USC 136
Federal Land Policy and Management Act of 1976, 43 USC 1701
Federal Lands Cleanup Act of 1985, 36 USC 169i
Federal Water Pollution Control Act Amendments of 1972, 33 USC 1251
Federal Water Project Recreation Act, 16 USC 4601
Fish and Wildlife Conservation Act of 1980, 16 USC 2901
Fish and Wildlife Coordination Act of 1958, 16 USC 661
Food Security Act of 1985, 16 USC 3811
Geothermal Energy Research, Development and Demonstration Act, 30 USC 1101
Global Climate Protection Act of 1987, 15 USC 2901
Hazardous Materials Transportation Act, 49 USC 1801
Hazardous Substance Response Revenue Act of 1980, 26 USC 4611
Historic Sites, Buildings, and Antiquities Act of 1935, 16 USC 461
Land and Water Conservation Fund Act, 16 USC 4601
Marine Mammal Protection Act of 1972, 16 USC 1361
Marine Plastic Pollution, Research and Control Act of 1987, 33 USC 1901
Marine Protection, Research and Sanctuaries Act of 1972, 33 USC 1401
Marine Resources and Engineering Development Act of 1966, 33 USC 1101

Migratory Bird Conservation Act, 16 USC 715
Migratory Bird Treaty Act, 16 USC 703
National Climate Program Act, 15 USC 2901
National Energy Conservation Policy Act, 42 USC 8201
National Environmental Policy Act of 1969, 42 USC 4321
National Historic Preservation Act of 1966, 15 USC 470
National Materials Policy Act of 1970, 42 USC 6901
National Ocean Pollution Planning Act of 1978, 33 USC 1701
National Trails Systems Act of 1986, 16 USC 1241
National Invasive Species Act of 1996, 16 USC 4701
Noise Control Act of 1972, 42 USC 4901
Noxious Plant Control Act, 43 USC 1241
Nuclear Waste Policy Act of 1982, 42 USC 10101
Occupational Safety and Health Act of 1970, 29 USC 651
Oil Pollution Act of 1990, 33 USC 2701
Outdoor Recreation Resources Review Act, 16 USC 17k
Outer Continental Shelf Lands Act Amendments of 1978, 43 USC 1801
Pollution Prevention Act of 1990, 42 USC 13101
Protection of Navigable Waters and of Harbors, 33 USC 401
Public Health Service Act, 42 USC 300f
Radon Gas and Indoor Air Quality Research Act of 1986, 42 USC 7401
Reservoir Salvage Act of 1960, 16 USC 469
Resource Conservation and Recovery Act of 1976, 42 USC 6901
River and Harbor Act of 1899, 33 USC 401
River and Harbor and Flood Control Act of 1970, 33 USC 426
Safe Drinking Water Act of 1974, 42 USC 300f
Soil and Water Resources Conservation Act of 1977, 16 USC 2001
Soil Conservation and Domestic Allotment Act, 16 USC 590a
Solid Waste Disposal Act, 42 USC 6901
Submerged Lands Act of 1953, 43 USC 1301
Superfund Amendments and Reauthorization Act of 1986, 42 USC 9601
Surface Mining Control and Reclamation Act of 1977, 30 USC 1201
Toxic Substances Control Act of 1976, 15 USC 2601
Wild and Scenic Rivers Act, 16 USC 1271

to submit environmental data that characterize the nature and extent of the discharge. A government permit writer develops a draft permit with discharge limitations, testing and monitoring, and other compliant conditions. Negotiations with the applicant may occur during this process and additional testing or other data may be required to support the permit application. The draft permit is made available for public comment. Following consideration of public comments, a final permit is issued by the regulatory authority. When monitoring is required, periodic data reports are required to be submitted to the regulatory authority, and these data are available to the public through the Freedom of Information Act procedures.

2.5 REGULATIONS

Regulations are final rules published in the *Federal Register* following an opportunity for public comment. They have the force and effect of law. The process may involve an Advanced Notice of Proposed Rule-Making (ANPRM) in which an agency proposes, for public comment, a general policy or regulatory thrust for the control of an activity. Following public comments, or directly if an ANPRM is not published, a Proposed Rule is developed and published in the *Federal Register* for public comment. The Proposed Rule contains the proposed regulatory language and a preamble that provides the applicable government contact, a summary of the rule-making procedure, and a discussion of the agency's intent in the development of each section of the proposed rule. In addition to the public comment period, the agency may hold one or more informal hearings on the proposed rule and solicit public statements. The agency must respond to the public comments received, and may make changes in the proposed regulatory language. These are addressed in the preamble to the final rule, which is developed and promulgated subsequently. See Table 2.5.

TABLE 2.5
Selected Significant Environmental Regulations

NEPA

40 CFR 6	EPA National Environmental Policy Act Procedures
40 CFR 1500	CEQ Regulations Implementing NEPA

Air

40 CFR 50	National Primary and Secondary Air Quality Standards
40 CFR 53	Ambient Air Monitoring Reference and Equivalent Methods
40 CFR 60	Standards of Performance for New Stationary Sources
40 CFR 61	National Emission Standards for Hazardous Air Pollutants
40 CFR 82	Protection of Stratospheric Ozone

Water

40 CFR 109	Criteria for State, Local, and Regional Oil Removal Contingency Plans
40 CFR 110	Discharge of Oil
40 CFR 112	Oil Pollution Prevention
40 CFR 113	Liability Limits for Small Onshore Oil Storage Facilities
40 CFR 116	Designation of Hazardous Substances
40 CFR 117	Determination of Reportable Quantities for Hazardous Substances
40 CFR 122	National Pollutant Discharge Elimination System Permit Regulations
40 CFR 123	State Program Requirements
40 CFR 125	Criteria and Standards for the National Pollutant Discharge Elimination System
40 CFR 129	Toxic Pollutant Effluent Standards
40 CFR 130	Requirements for Water Quality Planning and Management (Water Quality Standards)
40 CFR 136	Guidelines Establishing Test Procedures for the Analysis of Pollutants
40 CFR 140	Marine Sanitation Device Standard
40 CFR 403	General Pretreatment Regulations for Existing and New Sources of Pollution
40 CFR 501	State Sludge Management Program Regulations

Drinking Water

40 CFR 141	National Primary Drinking Water Regulations

40 CFR 142	National Primary Drinking Water Regulations Implementation
40 CFR 143	National Secondary Drinking Water Regulations
40 CFR 144	Underground Injection Control Program
40 CFR 146	UIC Program: Criteria and Standards

Ocean Dumping

| 40 CFR 220–229 | Ocean Dumping |

Dredge and Fill (Wetlands)

| 40 CFR 230 | Interim Regulations on Discharge of Dredged or Fill Material into Navigable Waters |
| 40 CFR 231 | Section 404 (c) Procedures |

Solid Wastes

40 CFR 240–241	Guidelines for the Thermal Processing of Solid Wastes and for the Land Disposal of Solid Wastes
40 CFR 243	Guidelines for Solid Waste Storage and Collection
40 CFR 244	Guidelines for Solid Waste Management of Beverage Containers
40 CFR 245	Guidelines for Resource Recovery Facilities
40 CFR 246	Guidelines for Source Separation for Materials Recovery
40 CFR 247	Guidelines for Procurement of Products That Contain Recycled Material
40 CFR 248	Guidelines for Federal Procurement of Building Insulation Products Containing Recovered Materials
40 CFR 249	Guidelines for Federal Procurement of Cement and Concrete Containing Fly Ash
40 CFR 250	Guidelines for Federal Procurement of Paper and Paper Products Containing Recovered Materials
40 CFR 252	Guidelines for Federal Procurement of Lubricating Oils Containing Rerefined Oil
40 CFR 255	Guidelines for Identification of Regions and Agencies for Solid Waste Management
40 CFR 257	Regulations on Criteria for Classification of Solid Waste Disposal Facilities and Practices
40 CFR 258	Criteria for Municipal Solid Waste Landfills

Hazardous Wastes

40 CFR 261	Identification and Listing of Hazardous Waste
40 CFR 262	Hazardous Waste Generators
40 CFR 263	Hazardous Waste Transporters
40 CFR 264	Regulations for Owners and Operators of Permitted Hazardous Waste Facilities
40 CFR 268	Regulations on Land Disposal Restrictions
40 CFR 270	Standards for Used Oil Processors and Refiners
40 CFR 279	Standards for the Management of Used Oil
40 CFR 280	Technical Standards and Corrective Action Requirements for Owners and Operators of Underground Storage Tanks

Superfund and EPCRA

40 CFR 300	National Oil and Hazardous Substances Contingency Plan
40 CFR 302	Designation, Reportable Quantities, and Notification Requirements for Hazardous Substances under CERCLA
40 CFR 355	Regulations for Emergency Planning and Notification under CERCLA
40 CFR 370	Hazardous Chemical Reporting and Community Right-to-Know Requirements

| 40 CFR 372 | Toxic Chemical Release Reporting Regulations |
| 40 CFR 373 | Regulations for Real Property Transactions under CERCLA |

TSCA

40 CFR 761	Polychlorinated Biphenyls Manufacturing, Processing, Distribution in Commerce, and Use Prohibitions
40 CFR 763	Asbestos
40 CFR 792	Good Laboratory Practice Standards
40 CFR 796	Chemical Fate Testing Guidelines
40 CFR 797	Environmental Effects Testing Guidelines
40 CFR 798	Health Effects Testing Guidelines
40 CFR 1515	Freedom of Information Act Procedures

Other Selected NON-40 USEFUL CFRs

7 CFR 658	Farm Land Protection Policy
15 CFR 930	Federal Consistency with Approved Coastal Management Programs
16 CFR 3501	Coastal Barrier Resources
29 CFR 1910.119	Process Safety Management of Hazardous Chemicals
29 CFR 1910.120	Hazardous Waste and Emergency Response
29 CFR 1910.1200	OSHA Hazard Communication Standard
32 CFR 97	Release of Classified Information
32 CFR 229	Protection of Archaeological Resources: Uniform Regulations
33 CFR 320–330	Clean Water Act Section 404 and Rivers and Harbor Act Section 10 Regulatory Programs
36 CFR 79	Curator of Federally-Owned and Administered Archaeological Collections
36 CFR 800	Protection of Historic and Cultural Properties
49 CFR 173	Shippers—General Requirements for Shipments and Packaging
49 CFR 171–179	Hazardous Materials Transportation Regulations
50 CFR 10, 18, 216, and 228	Regulations Concerning Marine Mammals
50 CFR 10.13	List of Migratory Birds
50 CFR 17.11	List of Endangered and Threatened Wildlife and Plants
50 CFR 402	Interagency Cooperation—Endangered Species Act

2.5.1   RISK ASSESSMENT

Risk assessment is the assessment of the relative risk to human health or the environment that may result from the introduction of a pollutant, a perturbation, or a governmental action related to environmental media.

3 Assessments and Impacts

3.1 THE NATIONAL ENVIRONMENTAL POLICY ACT OF 1969

The National Environmental Policy Act of 1969 (NEPA), implemented by 40 CFR 1500–1508, requires that federal agencies consider the potential significant environmental impacts of each major federal action before commencing the action. The NEPA process should be integrated with other planning activities at the earliest possible time to ensure that decisions reflect environmental values and avoid potential conflicts (40 CFR 1501.2). It involves three levels of potential documentation:

1. Categorical Exclusion (CATEX), which consists of a category of actions that do not individually or cumulatively have a significant effect on the human environment, and which have been found to have no such effect by the responsible federal agency. Neither an environmental assessment (EA) nor an environmental impact statement (EIS) needs to be prepared, and public involvement is not required. Each federal agency maintains their own set of CATEXs, which are proposed and made final through the rule-making process. When used for a particular project, each CATEX must be justified for its use in a memorandum for the files, which is signed by the appropriate authority and retained.
2. The Environmental Assessment is a document that addresses the need for the project, the characteristics of the proposed project, alternatives to the proposed project that were considered, the existing environment, potential impacts on the existing environment, and mitigation measures that can and cannot be implemented to neutralize adverse impacts. The EA is based on data already available. The heart of an EA is the presentation of environmental impacts of the proposed action and alternatives in comparative form (40 CFR 1502.14). An EA results in a finding of no significant impact (FONSI) or a conclusion of significant environmental impact with a recommendation that an EIS be prepared. The FONSI must be made available to the public (40 CFR 1501.4 (e)), but depending upon the geographical extent of the federal action, the public notification may vary from publication in local newspapers to publication in the *Federal Register* (40 CFR 1506.6 (b) (3)).
3. The Environmental Impact Statement is a document that addresses topics similar to those in an EA, but is more extensive in its presentation and may include data generated during the impact study. An EIS is required if a significant environmental impact is expected. An EIS requires *Federal Register* publication of notice of intent to prepare an EIS; public scoping meeting(s); draft EIS available and distributed for public comment; final

EIS that addresses public comments; and formal Record of Decision published in the *Federal Register.*

NEPA requires that all federal agencies shall "include in every recommendation or report on proposals for legislation and other major federal actions significantly affecting the quality of the human environment, a detailed statement by the responsible official on the environmental impact of the proposed action; any adverse environmental effects which cannot be avoided should the proposal be implemented; alternatives to the proposed action; the relationship between local short-term uses of man's environment and the maintenance and enhancement of long-term productivity; and any irreversible and irretrievable commitments of resources which would be involved in the proposed action should it be implemented."

Normally confined to about 150 pages without the appendices, the format for an EIS is recommended at 40 CFR 1502.10. See Table 3.1.

Information to be included in an EIS when discussing potential effects on the affected environment is shown in Table 3.2.

The potential impacts of the proposed action on the above subjects should be discussed. Where there is a potential for impact, mitigation measures that will be undertaken if the proposed action is implemented should be thoroughly discussed.

The stated purposes of an EIS are to serve as an action-forcing document to ensure that the policies and goals of the NEPA process are infused into the ongoing programs and actions of the federal government, to provide full and fair discussion of significant environmental impacts, and to inform decision makers and the public of the reasonable alternatives that would avoid or minimize adverse impacts or enhance the quality of the human environment.

Under Section 309 of the Clean Air Act, the Administrator of the EPA is directed to review and comment publicly on the environmental impacts of federal activities, including actions for which environmental impact statements are prepared. If after this review the Administrator determines that the matter is "unsatisfactory from the standpoint of public health or welfare or environmental quality," Section 309 directs the matter be referred to the Council on Environmental Quality (40 CFR 1504.1(b)).

TABLE 3.1
Format for an Environmental Impact Statement

Cover sheet
Summary
Table of Contents
Purpose of and need for action
Alternatives considered, including proposed action
Affected environment
Environmental consequences
Mitigation actions
List of preparers
List of agencies, organizations, and persons to whom copies of the statement are sent
Appendices (if any)

TABLE 3.2
EIS Information Related to Effects on the Affected Environment

Climate
Soils
Geology
Seismicity
Floodplains
Wetlands
Wildlife, including fish and fisheries and upland wildlife
Water quality
Air quality
Hazards and hazardous waste sites
Noise
Transportation and traffic
Land use
Waste disposal
Endangered and threatened species
Recreation
Historical and cultural resources
Population
Socioeconomic considerations

The EA or EIS should contain a statement regarding any disproportionate adverse health or safety impacts to low income or minority populations pursuant to Executive Order 12898.

3.2   FARMLAND PROTECTION POLICY ACT OF 1991

If the federal action impacts farmland, the provisions of the Farmland Protection Policy Act must be addressed in the NEPA documentation. Section 1539 of the Act has as its purpose to minimize the extent to which federal programs contribute to the unnecessary and irreversible conversion of farmland to non-agricultural uses, and to ensure that federal programs are administered in a manner that, to the extent practicable, will be compatible with state, local government, and private programs and policies to protect farmland.

Pursuant to the Act, the Soil Conservation Service promulgated 7 CFR 658. The preamble to this rule states, "Neither the Act nor this rule requires a federal agency to modify any project solely to avoid or minimize the effects of conversion of farmland to non-agricultural uses. The Act requires that before taking or approving any action that would result in conversion of farmland as defined in the Act, the agency examines the effects of the action, using the criteria set forth in the rule, and if there are adverse effects, consider alternatives to lessen them. The agency would still have discretion to proceed with a project that would convert farmland to non-agricultural uses once the examination required by the Act has been completed."

3.3   ENVIRONMENTAL EFFECTS ABROAD OF MAJOR FEDERAL ACTIONS

Executive Order 12114 on Environmental Effects Abroad of Major Federal Actions extends NEPA to overseas actions. The jurisdiction of NEPA, per se, is limited to the geographical boundaries of the United States. The requirements of EO 12114 are slightly different from those of NEPA. The types of assessment documents applicable to the Department of Defense actions are delineated below:

1. An Overseas Categorical Exclusion (OCE) is one that provides a factual summary to demonstrate that no exceptions to the OCE are applicable, sets out the applicable exemption on which it is based, succinctly states the decision to forego the preparation of an assessment document, and is approved by the commanding officer. It does not require further analysis or action.

2. An Overseas Environmental Assessment (OEA) is prepared unilaterally. It is an internal Department of Defense (DOD) document that does not require public participation; however, it is available upon request under the provisions of the Freedom of Information Act. It normally should not exceed 35 pages.

3. An Overseas Environmental Impact Statement (OEIS) is prepared for actions affecting the Global Commons and the United States Exclusive Economic Zone, and may result from the recommendation of an overseas environmental assessment. The Global Commons are geographical areas that are outside the jurisdiction of any nation, and include the oceans outside territorial limits and Antarctica. They are prepared unilaterally by the United States, but may be made available to foreign governments after coordination through the chain of command with the Department of State. Draft OEIS should not exceed 100 pages. Draft copy is made available to the public for comment.

4. Environmental Reviews (ER) are prepared for major federal actions that significantly harm the environment by introducing:
    a. A prohibited or strictly regulated toxic product, effluent, or emission, such as asbestos, vinyl chloride, PCBs, mercury, beryllium, or arsenic or
    b. A physical project that in the United States is prohibited or strictly regulated by federal law to protect the environment against radioactive substances. An environmental review is prepared unilaterally and does not involve formal contact or consultation with the host nation.

5. An Environmental Study (ES) is a cooperative bilateral or multilateral study prepared with one or more foreign nations. Careful coordination with the Department of State is required by the major claimant. Generally, it is more detailed than an ER and usually it is 10 to 50 pages in length.

4 Water

Conditions that led to legislative actions in the 1970s were those seen by the individual senators and representatives or reported to them through their constituents. Such conditions were those discussed below.

In the decades of the 1950s and 1960s, many waterways in the United States showed signs of gross pollution. The Potomac River tidal basin near the nation's capital was too dirty to provide swimming; the Cuyahoga River burned because of the oil on its surface; and Lake Erie was declared dead because there was no oxygen to support fish in its deeper waters. The Klamath River in Oregon and Washington was still being used to float logs to a papermill, and at times navigation was halted thereby; the city of St. Louis collected its garbage, ground it, and then barged it to the center of the Mississippi River for dumping; and the Flambeau River in Wisconsin carried so much black liquor from a sulphite papermill that its bottom was covered with wood fibers, the river would rumble from escaping gases of decomposition, and boils of wood fibers and slime would rise to the river surface 6 in. thick and 10 ft in diameter. Many lakes of that era were overcome with algal and other vegetation growth from excessive nitrogen and phosphorus nutrients. These lakes included Lake Sebasticook in Maine, Lake Okeechobee in Florida, the Madison Lakes in Wisconsin, Detroit Lakes in Minnesota, and Lake Washington in Washington. Boston Harbor was also severely polluted because of a marine alagal growth caused by the nutrients in primary-treated sewage being discharged through an ocean outfall.

In the early days, the mayfly emergence (*Hexagenia sp.*), locally known as the Green Bay fly from Green Bay, WI, was so great that road graders had to be dispatched to clear bridges and roadways of the bodies of dead adult mayflies. This phenomenon has been described from the literature (Mackenthun, 1969) for the Mississippi River, "Unusual hordes of these insects may leave the water on the first suitable day for hatching after a period of adverse weather. The adults are fragile insects that die within a few hours. When occurring in hordes, their dead bodies may clog ventilator ducts and sewers and may cause temporary traffic difficulties. On July 23, 1940, at Sterling, IL, mayflies piled as high as 4 ft and blocked traffic over the Fulton-Clinton highway bridge for nearly 2 h. Fifteen men in hip boots used shovels and a snowplow to clear a path."

The Fox River, draining into Green Bay in Wisconsin supports several papermills. The process used to convert wood fibers to paper in the 1940s produced a waste that had a very high biochemical oxygen demand and removed the dissolved oxygen when the waste was introduced into receiving water.

The fish gradually disappeared from the lower Fox River. The Fox River water tended to hug the bottom and eastern shore of Green Bay. As the water, devoid of dissolved oxygen, worked its way up the eastern shore, the bountiful supply of immature mayflies left their normal habitat in the bottom sediments because of a lack of oxygen. The immature mayflies began to cling to commercial yellow perch fishing nets, which were set off the bottom in areas where some dissolved oxygen still remained. The immature mayflies resembled clusters of grapes clinging to the fishermen's nets and the nets became useless as fishing tools. As time advanced, dissolved oxygen in the lower Green Bay disappeared and became non-existent. The once ponderous mayfly population gradually disappeared. In 1953, the author collected over 100 bottom samples from Green Bay through the ice. Only one immature mayfly was found in these samples and was captured near the western Green Bay shore.

4.1   SURFACE FRESH WATER

1. *Evolution of the Clean Water Act*—The oldest federal environmental law, the Clean Water Act, evolved over the years to its present state. It began with the River and Harbor Act of 1899. The principal sequential events to the present are listed below:
   a. *1899—River and Harbor Act*—The 1899 law is still viable. Section 13 of the River and Harbor Act of 1899 made it unlawful to put any refuse matter into navigable waters other than that flowing from streets and sewers in a liquid state, except by permit from the Corps of Engineers. Section 16 provided that half of the fine following violation shall go to the person providing information leading to conviction.
   b. *1912—Public Health Service Act*—Provided for the investigation of water pollution in relation to diseases and impairment of humans.
   c. *1924—Oil Pollution Act*—Made it unlawful to discharge oil into the coastal navigable waters, subject to a maximum fine of $2,500.
   d. *1948—Water Pollution Control Act*—placed the federal program in the Public Health Service of the Federal Security Agency, provided for federal loans to construct sewage treatment works, and authorized the formation of interstate compacts to control water pollution.
   e. *1956—Pub L-660*—Provided for state program grants and construction grants for treatment works. The federal government was provided enforcement authority for water pollution control for interstate waters.
   f. *1961—Amendments*—Provided for federal enforcement of water pollution in interstate and navigable waters.
   g. *1965—Water Quality Act*—Created state water quality standards and water use designations for navigable waters.
   h. *1966—Reorganization Plan #2*—Transferred the federal water pollution control program to the Department of the Interior.
   i. *1970—Executive Order 11574 of December 23, 1970*—Implemented a federal permit program for Section 13 of the River and Harbor Act of 1899 to be administered by the Corps of Engineers.
   j. *1970—Reorganization Plan #3*—Established the Environmental Protection Agency.

k. *1972—Pub L 92-500*—Created by Senate Bill 2770 and House Bill 9560. The Federal Water Pollution Control Act provided a national comprehensive water pollution control program.

l. *1976*—Following lawsuits by environmental groups against the EPA for not implementing effluent guidelines in a timely fashion, the resultant Settlement Agreement signed by the EPA, Natural Resources Defense Council, Environmental Defense Fund, and Citizens for a Better Environment, provided a schedule for developing effluent guidelines for 27 industrial categories to address 65 priority pollutants, and for the development of water quality criteria for the same pollutants.

m. *1977*—Amendments to the Federal Water Pollution Control Act virtually wrote into law the Settlement Agreement provisions for toxic pollutant control; it shortened the name of the law to the Clean Water Act.

n. *1987—Clean Water Act Amendments*—Established Section 402 (p) to control stormwater discharges through permits, and required toxic pollutant criteria to be included in all state water quality standards.

2. Titles of the Clean Water Act include:
   a. Research and Related Programs
   b. Grants for Treatment Works
   c. Standards and Enforcement
   d. Permits and Licenses
   e. General Provisions
   f. State Water Pollution Control Revolving Funds

   The titles of greatest interest to this discussion are titles c and d. Title c provides the legislative requirements for controlling pollution and title d provides the permit mechanism for affecting such controls. Title a, however, provides the Act's objective, "It is the objective of this Act to restore and maintain the chemical, physical, and biological integrity of the Nation's waters."

3. The Clean Water Act attacks pollution from the water quality aspect and from the permit treatment technology aspect. We will begin with the water quality provisions. Selected sections of the Clean Water Act are discussed below:

   a. Section 101 provides the goals of the Clean Water Act which are that the discharge of pollutants into the navigable waters be eliminated and that, wherever attainable, the water quality be such as to support fishing and swimming.

   b. Section 303 mandates that states develop water quality standards to protect the following designated water uses:
   Recreation and aesthetics
   Public water supply
   Fish and wildlife
   Agriculture
   Industry

   Once established, water quality standards are reviewed by states at public hearings every three years. Such standards designate water uses for each waterway within a state, provide water quality criteria to support such uses, include an antidegradation policy, and detail a means of enforcing the antidegradation policy.

c. Section 304 mandates that EPA develop and issue federal water quality criteria. A selected history of water quality criteria development includes:

1917 Shelford—published data concerning the effects on fish for a large number of gas plant constituent wastes.

1937 Ellis—reviewed the existing literature for 114 substances and in a 72-page document for the U.S. Bureau of Fisheries listed lethal concentrations found by the various authors. He provided a rationale for the use of standard test animals in aquatic bioassay procedures and used the goldfish, *Carassius auratus,* and the entomostracan, *Daphnia magna,* as test species on which experiments were made in constant temperature cabinets.

1952 The State of California published a 512-page book on Water Quality Criteria that contained 1,369 references. This classic reference summarized water quality criteria promulgated by state and interstate agencies, as well as the legal application of such criteria.

1963 The State of California published a completely revised Water Quality Criteria book edited by McKee and Wolf, which included 3827 references. This was a monumental effort in bringing together under one cover the world's literature on water quality criteria.

1968 The Secretary of the Interior appointed several nationally recognized scientists to a National Technical Advisory Committee to develop water quality criteria for five specified uses of water: agricultural, industrial, recreational, fish and wildlife, and domestic water supply. This report, known as the Green Book because of the color of its cover, constituted the most comprehensive documentation to date on water quality requirements for particular and defined water uses.

1974 The Blue Book resulted from a contract from EPA to the National Academy of Sciences and the National Academy of Engineering to expand the concept of the Green Book and to develop a water quality criteria document that would include current knowledge.

1976 Quality Criteria for Water, called the Red Book, was developed within the EPA and was the first water criteria document published as required by Section 304(a) of the Clean Water Act.

1980 The EPA published 65 separate criteria documents for the 65 toxic pollutants.

1986 The Gold Book, Quality Criteria for Water, is the latest EPA criteria document. In essence, it represents a summary of the 1980 65 criteria documents combined with some of the 1976 criteria.

d. Sections 306 and 307 (b) require the development of effluent guidelines to provide national standards of performance for industrial discharges and pretreatment standards for industries that discharge to publicly-owned treatment works (POTW).

e. Section 307 (a) requires a list of toxic pollutants. The 1977 Amendments, by Act of Congress, renamed the 65 priority pollutants as toxic pollutants. See Table 4.1.

TABLE 4.1
The 65 Toxic Water Pollutants

1. Acenaphthene
2. Acrolein
3. Acrylonitrile
4. Aldrin/Dieldrin
5. Antimony and compounds
6. Arsenic and compounds
7. Asbestos
8. Benzene
9. Benzidine
10. Beryllium and compounds
11. Cadmium and compounds
12. Carbon tetrachloride
13. Chlordane (technical mixture and metabolites)
14. Chlorinated benzenes (other than dichlorobenzenes)
15. Chlorinated ethanes (including 1,2-dichloroethane, 1,1,1-trichloroethane, and hexa-chloroethane)
16. Chloroalkyl ethers; chloroethyl and mixed ethers
17. Chlorinated naphthalene
18. Chlorinated phenols (other than those listed elsewhere; including trichlorophenols and chlori-nated cresols)
19. Chloroform
20. 2-Chlorophenol
21. Chromium and compounds
22. Copper and compounds
23. Cyanides
24. DDT and metabolites
25. Dichlorobenzenes (1,2-, 1,3-, and 1,4-dichlorobenzenes)
26. Dichlorobenzidine
27. Dichloroethylenes (1,1- and 1,2-dichloroethylene)
28. 2,4-Dichlorophenol
29. Dichloropropane and dichloropropene
30. 2,4-Dimethylphenol
31. Dinitrotoluene
32. Diphenylhydrazine
33. Endosulfan and metabolites
34. Endrin and metabolites
35. Ethylbenzene

36. Fluoranthene
37. Haloethers (other than those listed elsewhere, including bromophenylphenyl ether, bis-dichloroisopropyl ether, bis-chloroethoxy methane and polychlorinated diphenyl ethers)
38. Halomethanes (other than those listed elsewhere; including methylene chloride, methylchloride, methylbromide, and bromoform, dichlorobromomethane)
39. Heptachlor and metabolites
40. Hexachlorobutadiene
41. Hexachlorocyclohexane
42. Hexachlorocyclopentadiene
43. Isophorone
44. Lead and compounds
45. Mercury and compounds
46. Naphthalene
47. Nickel and compounds
48. Nitrobenzene
49. Nitrophenols (including 2,4-dinitrophenol, dinitrocresol)
50. Nitrosamines
51. Pentachlorophenol
52. Phenol
53. Phthalate esters
54. Polychlorinated biphenyls (PCBs)
55. Polynuclear aromatic hydrocarbons (including benzanthracenes, benzopyrenes, benzofluoranthene, chrysenes, dibenzathracenes, and indenopyrenes)
56. Selenium and compounds
57. Silver and compounds
58. 2,3,7,8-Tetrachlorodibenzo-p-dioxin (TCDD)
59. Tetrachloroethylene
60. Thallium and compounds
61. Toluene
62. Toxaphene
63. Trichloroethylene
64. Vinyl chloride
65. Zinc and compounds

---

f. Section 308 of the Clean Water Act provides a means of collecting data and reports from the industries for which an effluent guideline is being developed. A letter written under the auspices of Section 308 that requests testing or information must, under penalty of law, be responded to by the recipient. Such a letter cannot be forwarded from EPA without first obtaining approval of the Office of Management and Budget in the President's office.

g. Section 311 requires the designation of hazardous substances with reportable quantities and provides penalties for the discharge of oil or hazardous substances. Hazardous substances are designated at 40 CFR 116. The Oil Pollution Act of 1990 amends Section 311 (j) of the Clean Water Act to require facility response plans for a worst case discharge of oil and to a substantial threat of such a discharge. The regulations require "sensitivity or vulnerability" analyses for environmentally sensitive areas. See Table 4.2.

TABLE 4.2
Sensitive Ecological and Structural Areas

Water intakes
Schools
Medical facilities
Residential areas
Businesses
Wetlands
Fish and wildlife
Lakes and streams
Threatened and endangered species
Recreational areas
Transportation routes
Utilities
Other areas of economic importance

---

h. Section 312 mandates promulgation of regulations to control the discharge of untreated or inadequately treated sewage from boats. For marine waters, a macerator-chlorinator marine sanitation device is adequate. For some inland lakes, sewage holding tanks are required. Section 312(n) provides for the development of Uniform National Discharge Standards for discharges incidental to the operation of vessels of the Armed Forces. Discharges incidental to vessel operation must be identified by February 10, 1998; standards of performance for the marine pollution control devices are to be developed by February 10, 2000; and regulations governing design, construction, installation, and use of the marine pollution control devices are to be promulgated by February 10, 2001.

i. Section 319 provides for a nonpoint source pollution control program, initially consisting of state developed plans for control.

j. Section 320 provides for a National Estuary Program, whereby conferences are held, studies are conducted, pollution control plans are developed, and decisions are made for implementing the control plans. Currently 12 estuaries are under comprehensive study.

k. Section 401 provides that any applicant for a federal license or permit for a discharge to the waters of the Unites States must obtain state certification that the license or permit will comply with the state water quality standards.

l. Section 402 provides for the National Pollutant Discharge Elimination System (NPDES) permitting process. The regulations implementing the permit program are found at 40 CFR Part 122. Section 301 (b) (1) (C) mandates that NPDES permits be developed to meet water quality standards. NPDES permits contain:
   • Discharge limitations based upon treatment technology effluent guidelines, best engineering judgment in the event effluent guidelines

are not available, or constituent concentrations that will not violate water quality standards if conditions require that the permit be based on water quality rather than treatment technology.

- Provisions for biological toxicity test if applicable. This requires subjecting test organisms, i.e., fish and an invertebrate, to different concentrations of the wastewater to determine the wastewater concentration that will be lethal to 50 percent of the test organisms within a 96-hour period.
- Monitoring requirement and submittal of discharge monitoring data to the regulatory authority on a periodic schedule.
- Other management practices to meet permit conditions.

m. Section 404 provides that the Corps of Engineers shall issue permits for the discharge of dredged or fill material into the waters of the United States. The issuance of such permits must take into consideration the disposal guidelines prepared by the EPA in conjunction with the Corps of Engineers. The section states, further, that the Corps may issue a permit, regardless of the guidelines, based upon economic considerations and navigability. However, Section 404(c) provides that the EPA may veto any such permit, after notice and public hearing, based upon a determination of environmental requirements. The Corps issues individual Section 404 permits for specific projects, or general permits. In FY94 there were over 48,000 applications for permits. General permits covered 82 percent of the application and were processed in an average of 16 days. See Table 4.3 for more information.

TABLE 4.3
Process for Developing a Regulation, Effluent Guideline, and NPDES Permit

Regulation	Effluent Guideline	Permit
• Program office develops scope and language	• Industrial category selection	• Wastewater sampling and analyses
• Development plan	• Development plan	• Permit application
	• Engage contractor— sampling and analyses	• Permit writer develops draft permit
	• Issue Section 308 letter	• Effluent guidelines;
	• Economic impact studies of alternative guidelines	best engineering judgment; water quality
• Work group	• Work group	based; toxicity testing
• Draft regulation	• Draft effluent guideline	• Negotiate with permittee
• Steering committee	• Steering committee	• Public comment on draft
• Red border signatures	• Red border signatures	permit
• Administrator signature	• Administrator signature	• Public hearing or
• Publish in *Federal Register*	• Publish in *Federal Register*	consider public comments
• Hearing	• Issue development document	• Develop administrative record
• Consider public comments	• Consider public comments	• Issue final permit

- Work group deliberation
- Steering committee
- Red border signatures
- Administrator signature
- *Federal Register*
- CFR codification

- Work group deliberation
- Steering committee
- Red border signatures
- Administrator signature
- *Federal Register*
- CFR codification

A brief review of the Clean Water Act is stimulated by Table 4.4. Section 301(a) requires a permit for the discharge of a pollutant into the navigable waters of the United States. To find out what is included within navigable waters, one must go to the definitions in Section 502. The definition of navigable waters, however, is quite brief in stating that navigable waters are waters of the United States. (A more comprehensive definition of waters of the United States may be found in 40 CFR 122.2) The regulatory aspects of the Clean Water Act then is divided into two parts: the permitting activity centered in Section 402, and the water quality activity centered in Section 303.

TABLE 4.4
Diagram Summarizing the CWA Regulatory Process

		Section 301 (a)		
		Section 502		
Section 304 (a) . . .	Section 303 . . .	Section 301 (b)	Section 402 . . .	Section 304 (b)
		(1) (C)		Section 306
			Section 401	Section 307 (a)
		Section 302		Section 307 (b)
		Section 311		Section 308
		Section 318		Section 309
		Section 404		

Section 301 (a)—unlawful to discharge pollutant without a permit.
Section 301 (b) (1) (C)—permits must not violate water quality standards.
Section 302—provides for water quality based permitting.
Section 303—establishes water quality standards program.
Section 304 (a)—provides for federal water quality criteria.
Section 304 (b)—effluent guidelines for existing sources.
Section 306—effluent guidelines for new sources.
Section 307 (a)—toxic pollutant list.
Section 307 (b)—pretreatment standards for waste treatment works.
Section 308—mandatory response information letter.
Section 309—federal enforcement provisions.
Section 311—oil and hazardous substance spills.
Section 312—sewage and incidental discharges from vessels.
Section 318—Aquaculture project discharge control.
Section 401—state certification provisions.
Section 402—National Pollutant Discharge Elimination System.
Section 404—Discharge of dredged and fill material.
Section 502—Clean Water Act definitions.

Several other sections of the Act support Section 402. These include the effluent guidelines activity developed in accordance with Sections 304 and 306; toxic pollutant restrictions in Section 307 (a); pretreatment requirements in Section 307 (b); and enforcement provisions in Section 309. Section 308 letters generally are used to complete the information activities for a Section 402 permit. Section 401 requires that any federal project that requires a license also requires state certification that water quality standards will not be violated because of its implementation.

Section 303, water quality standards, is designed to protect water quality for identified uses of the waterway. The water quality standards are supported by Section 304 (a), the federal water quality criteria program. The permit process for point source discharges and water quality standards are linked together by Section 301 (b) (1) (C), which states that any National Pollutant Discharge Elimination System permit issued pursuant to Section 402 must meet the provisions of Section 303 water quality standards.

President Clinton, announced a Clean Water Action Plan containing four tools to achieve clean water goals and clean the waters that are still polluted (63 FR 14109, March 24, 1998). These tools are a watershed approach—the key to the future, strong federal and state standards, natural resource stewardship, and informed citizens and officials. Among the goals is a promise that EPA will establish by the year 2000 numeric criteria for nitrogen and phosphorus nutrients for lakes, rivers, and estuaries. In 1968, the writer published a considered judgment, suggesting that to prevent biological nuisances, total phosphorus should not exceed 100 $\mu g/l$ of P at any point within the flowing stream, nor should 50 $\mu g/l$ of P be exceeded where waters enter a lake, reservoir, or other standing water body. Those waters now containing less phosphorus should not be allowed to become phosphorus degraded. It will be interesting to see how closely that criterion will resemble the one promised for the year 2000.

4.2   NON-INDIGENOUS AQUATIC NUISANCE PREVENTION AND CONTROL ACT OF 1990

The Congress, in developing this Act, found that the discharge of untreated water in the ballast tanks of vessels and through other means results in the unintentional introduction of nonindigenous species to fresh, brackish, and saltwater environments. When environmental conditions are favorable, non-indigenous species, such as the zebra mussel, *Dreissena polymorpha,* become established and may disrupt the aquatic environment and economy of affected coastal areas. The Congress found that the potential economic disruption to communities affected by the zebra mussel owing to its colonization of water pipes, boat hulls, and other hard surfaces has been estimated to be $5 billion by the year 2000.

This Act required the U.S. Coast Guard to issue voluntary guidelines to prevent the introduction and spread of aquatic nuisance species into the Great Lakes through the exchange of ballast water of vessels prior to entering those waters. Within 24 months, regulations are to be issued.

4.3   NATIONAL INVASIVE SPECIES ACT OF 1996

The National Invasive Species Act of 1996 amended the Non-indigenous Aquatic Nuisance Prevention and Control Act to prevent the introduction and spread of non-indigenous species into the waters of the United States. A ballast management practice known as high seas ballast exchange greatly reduces the transfers of dangerous organisms through ballast water. In this amendment, the Congress found that an estimated 21 billion gal of ballast water from vessels from foreign ports is discharged into U.S. waters each year. It is estimated that 3,000 species of aquatic organisms are in transit in ballast tanks around the world in any given 24 h period. Over 3 billion gal of ballast water a year is discharged into the Chesapeake Bay from ships calling at the ports of Baltimore and Norfolk. This water originates from as least 48 foreign ports. An ongoing study by the Smithsonian Environmental Research Center found that nearly 90 percent of the vessels had living organisms in their ballast water.

The 1990 Act was designed to protect the Great Lakes from the further invasion on non-indigenous species. The 1996 Act was designed to protect all territorial waters of the United States. The Act required the U.S. Coast Guard to issue voluntary ballast water management guidelines with mandatory reporting by each affected vessel to the U.S. Coast Guard. The ballast water exchange is to be instituted when entering the 200 mi Exclusive Economic Zone of the United States. If the voluntary program is found not to provide adequate control of invasive species, the program is to be made mandatory within two years.

4.4   GROUND WATER

4.4.1   SAFE DRINKING WATER ACT OF 1974

Certain provisions of the Safe Drinking Water Act protect the ground water aquifers. In addition, this Act provides standards to ensure safe drinking water at the tap.

1. Makes drinking water standards applicable to all public water systems with at least 15 service connections serving at least 25 individuals.
   a. Drinking water standards were first adopted in 1914, revised in 1942 and 1962; but were applicable *only* to interstate carriers, such as railroads, buses, and airplanes, prior to 1974.
   b. Requires promulgation of primary drinking water regulations that specify maximum contaminant levels *at the tap* for constituents that may have any adverse effect on the health of persons.
   c. Provides for secondary drinking water regulations, which are *unenforceable at the federal level,* that specify maximum contaminant levels to protect public welfare, principally infrastructures, equipment, and against iron deposits when white clothing is washed.
   d. No national primary drinking water regulation may require the addition of any substance for preventive health care purposes unrelated to the contamination of drinking water.
   e. Prohibits the uses of pipe, solder, fittings, and flux that is not lead free in public water systems.
   f. Provides for the protection of underground sources of drinking water through issuance of regulations for state underground injection programs, sole-source aquifer protection where the vulnerability of an aquifer is owing to hydrogeologic characteristics, and a wellhead protection program.

2. Approximately 51 percent of the public drink groundwater, 49 percent drink surface water from rivers and reservoirs.
3. National Interim Primary Drinking Water Regulations protect health to the extent feasible, using technology, treatment techniques, and other means, and taking costs into consideration. Either maximum contaminant levels or treatment techniques requirements for the contaminants regulated may be prescribed.
4. Regulations:
   - 40 CFR 141—National Interim Primary Drinking Water Regulations.
   - 40 CFR 142—National Interim Primary Drinking Water Regulations Implementation.
   - 40 CFR 143—National Secondary Drinking Water Regulations.
   - 40 CFR 144—Underground Injection Control (UIC) Program.

   Class I wells: Used by generators of hazardous waste management facilities to inject hazardous waste beneath the lowermost formation containing, within 1/4 mile of the well bore, an underground source of drinking water.

   Class II wells: Injection from conventional oil or natural gas production for the enhanced recovery of oil or natural gas.

   Class III wells: Injection for solution mining activities (sulfur, salts, and potash).

   Class IV wells: Disposal of hazardous or radioactive wastes within 1/4 mile of a well that contains an underground source of drinking water.

   Class V wells: Other.
   - 40 CFR 145—State UIC Programs.
   - 40 CFR 146—UIC Program Criteria and Standards.
   - 40 CFR 147—State UIC Programs (Individual States).
5. Treatment of incoming water:
   a. pH and alkalinity adjustment: High and low pH levels accelerate metals corrosion.
   b. Coagulation with aluminum sulfate and sedimentation.
   c. Filtration:
      - Slow sand filters—0.05 gal/min/ft^2 of filter area.
      - Rapid sand filters—2.0 gal/min/ft^2 of filter area.
      - Diatomaceous earth filter.
   d. Aeration to remove volatile substances.
   e. Activated carbon for taste and odor control, and polishing.
6. Organisms potentially in water supplies:
   a. Taste and odor algae.
   b. Filter clogging algae.
   c. Iron bacteria in wells.
   d. Animals such as bloodworms, Nais worms, nematodes, copepods, clams, and snails.
   e. Parasites such as *Cryptosporidium.*

7. The Safe Drinking Water Act mandates the EPA to promulgate:
   a. Maximum Contaminant Level Goals (MCLGs), which are the concentrations at which no known or anticipated adverse effects on public health will occur, allowing for an adequate margin of safety.
   b. Maximum Contaminant Levels (MCLs), which represent the concentrations that cannot be exceeded in the treated water delivered to customers. MCLs must be set as close to the MCLG, as feasible with consideration for detection capabilities of laboratory test methods, treatment technologies and associated costs, and the economic burden placed on water suppliers.

4.5 MARINE WATER

4.5.1 MARINE PROTECTION, RESEARCH, AND SANCTUARIES ACT OF 1972 (MPRSA)

The marine waters are protected by the Ocean Dumping Act (MPRSA), the International Convention for the Prevention of Pollution from Ships (MARPOL), and the Act to Prevent Pollution from Ships. These Acts will be discussed in the preceding order. First, MPRSA:

1. Provides that the EPA may issue permits for the transportation from the United States of material for the purpose of dumping it into ocean waters.
   a. Radiological, chemical, and biological warfare agents, and high-level radioactive wastes are prohibited from permitting.
   b. Sewage sludge or industrial wastes were prohibited from ocean disposal as of December 31, 1991 by the Ocean Dumping Ban Act of 1988.
   c. MPRSA does not apply to the intentional placement of any device pursuant to an authorized federal or state program.
2. Provides that the Corps of Engineers may issue permits for the transportation of dredged material for the purpose of dumping it into ocean waters.
   a. The Secretary of the Army must first notify the Administrator of the EPA of the intent to issue a permit.
   b. In case of disagreement concerning the permit, the determination of the Administrator shall prevail, except that if there is no other economically feasible method or site available, the Secretary may request a waiver. The Administrator must grant the waiver within 30 days unless he finds that dumping will result in an unacceptable adverse environmental impact.
3. No permit shall be issued for the dumping of a material that will violate applicable water quality standards.
4. The following factors must be considered in evaluating a permit application:
   a. The need for the proposed dumping.
   b. The effect on human health and the environment.

    c. The effect on fisheries resources, shellfish, wildlife, shorelines, and beaches.

    d. The effect on marine ecosystems.

    e. The persistence and permanence of any environmental effects.

    f. The effects of volumes and concentrations of the material to be dumped.

5. Regulations:

    40 CFR 220—Provides for general permits, special permits, emergency permits, research permits, and permits for incineration at sea.

    40 CFR 221—Provides permit application provisions.

    40 CFR 222—Describes action on permit application.

    40 CFR 223—Describes permit contents.

    40 CFR 224—Mandates records and reports.

    40 CFR 225—Provides criteria for evaluating permits.

    40 CFR 227—Discusses uses of bioassays; limitations on disposal rates; waste quantities; hazards to fishing, navigation, shorelines or beaches; and criteria for dredged spoil disposal.

    40 CFR 227—Provides the basis for determining need for ocean dumping and the basis for determining impacts on ocean uses.

    40 CFR 227—Uses applicable marine water quality criteria or 96 h bioassay results to determine limiting permissible concentration of pollutant after mixing.

    40 CFR 227—Defines complete mixing as occurring after 4 h in an area bounded by the release zone and extending to the ocean floor or 20 m, whichever is shallower. The release zone is defined as extending 100 m from the perimeter of the conveyance engaged in the dumping action.

    40 CFR 228—Provides for designation of disposal sites.
- 40 CFR 228.6 (a) (11) requires an EIS for each site designation; proposed EIS for site proposal, final EIS for final rule.
- Provides listing of all designated dumping sites with coordinates.

    40 CFR 229—Provides conditions for General Permit.
- 40 CFR 229.1—Burial at sea.
- 40 CFR 229.2—Transport of target vessels.
- 40 CFR 229.3—Transportation and disposal of vessels.

The Ocean Dumping Ban Act of 1988 contains a Title III, "Dumping of Medical Wastes," which amends both the Clean Water Act and the Ocean Dumping Act to prohibit the dumping of medical wastes in fresh waters and nearer than 50 nautical miles (nm) from land in marine waters.

4.6 CONVENTION FOR THE PREVENTION OF POLLUTION FROM SHIPS (MARPOL)

Article 3 (3) states that the present Convention shall not apply to any warship, naval auxiliary, or other ship owned or operated by a state and used, for the time being, only on government non-commercial service. However, each Party shall ensure by the adoption of appropriate measures that all ships act in a manner consistent with MAR-POL, so far as is reasonable and practical.

4.6.1 ANNEX 1—PREVENTION OF POLLUTION BY OIL

1. Oceanic water, excluding Special Areas:
   a. Effective July 6, 1998, the effluent oil content without dilution shall not exceed 15 ppm when not in a Special Area.
   b. For ships not equipped with oil filtering equipment and an alarm to ensure the ship does not exceed a discharge of 15 ppm oil, the oil discharge standard is 100 ppm until July 6, 1998 or the date on which the ship is fitted with such equipment, whichever is earlier.
2. Within Special Areas:
   a. A ship of 400 tons gross tonnage and above shall retain on board all oil drainage and sludge, dirty ballast, and tank washing waters, and discharge them only to reception facilities.
   b. A ship of less than 400 tons gross tonnage shall meet an effluent oil content without dilution that does not exceed 15 ppm.
3. MARPOL Special Areas in effect include the Mediterranean Sea, Black Sea, and Baltic Sea. The Antarctic Area is in effect as of November 20, 1995, and there shall be no discharge of oily residues either ashore or into the sea. All waste should be removed from the Special Area. Special Areas that have been designated, but are not yet in effect, are the Red Sea, Persian Gulf, and the Gulf of Aden between the Red Sea and the Arabian Sea. The Gulf of Aden will become a special area, not in effect, on November 20, 1995.

4.6.2 ANNEX 2—PREVENTION OF BULK HAZMAT POLLUTION

This provides that Category A substances are prohibited from discharge; Category B substances are prohibited except when meeting described restrictions; Category C substances are prohibited except when not exceeding 10 ppm in the vessel wake; and Category D substances are prohibited except when concentration is less than 1 part in 10.

4.6.3 ANNEX 3—PREVENTION OF CONTAINERIZED HAZMAT POLLUTION

This provides restrictions for the spilling of containerized hazardous material, and requires that each Party to the Convention issue detailed requirements on packaging, marking, and labelling, documentation, stowage, quantity limitations, exceptions, and notification for preventing or minimizing pollution of the marine environment by harmful substances.

4.6.4 ANNEX 4—PREVENTION OF POLLUTION BY SEWAGE

This has not yet been ratified by the United States. The restrictions include:

1. No discharge within 4 nm of land.
2. Comminuted and disinfected sewage discharge allowed greater than 4 nm from land.
3. Untreated sewage discharge allowed greater than 12 nm from land provided discharge is at a moderate rate from a holding tank.

4.6.5 ANNEX 5—PREVENTION OF POLLUTION OF GARBAGE

1. Outside Special Areas:
   a. The discharge of all plastics into the sea is prohibited.
   b. Allows disposal of comminuted or ground food wastes, paper products, rags, glass, metal, bottles, and crockery that will pass through a 25 mm screen greater than 3 nm from land.
   c. Allows disposal of food wastes, paper products, rags, glass, metal, bottles, and crockery greater than 12 nm from land.
   d. Allows discharge of floatable dunnage, lining, and packing materials greater than 25 nm from land.
2. Special Areas:
   a. Special Areas in effect include the Baltic Sea, North Sea, and the Antarctic Area. Special Areas that have been designated but which are not in effect include the Mediterranean Sea, Black Sea, Red Sea, Persian Gulf, and the Wider Caribbean region, including the Gulf of Mexico.
   b. The discharge of all plastics is prohibited.
   c. The discharge of paper products, rags, glass, metal, bottles, crockery, dunnage, lining, and packing materials is prohibited.
   d. Allows the discharge of food wastes greater than 12 nm from land.
   e. In the Wider Caribbean region only, the discharge of comminuted or ground victual wastes that pass a 25 mm screen would be allowed greater than 3 nm from land, when the Special Area becomes effective.

4.7 ACT TO PREVENT POLLUTION FROM SHIPS (APPS) OF 1980

1. The Act was amended in 1987, 1993, and 1996.
2. The Act provides that MARPOL Annex 5 shall apply to the navigable waters of the United States, as well as to all other waters and vessels over which the United States has jurisdiction.
   a. Applicable to all ships after December 31, 1993, except those owned or operated by the Department of the Navy.
   b. Applicable to Navy surface ships after December 31, 1998.
   c. Applicable to Navy submarines after December 31, 2008.
   d. Navy surface ships are required to meet all Annex 5 Special Area provisions by December 31, 2000.
   e. Navy submarines are required to meet all Annex 5 Special Area provisions by December 31, 2008.

3. Notwithstanding any effective dates, the disposal of plastic shall apply to ships equipped with plastic processors upon the installation of such processors in such ships.
4. Except when necessary for the purpose of securing the safety of the ship, the health of the ship's personnel, or saving life at sea, it shall be a violation:
   a. With regard to a submersible, to discharge buoyant garbage or garbage that contains more than the minimum amount practicable of plastic.
   b. With regard to a surface ship, to discharge plastic contaminated by food during the last 3 days before the ship enters port.
   c. With regard to a surface ship, to discharge plastic, except plastic that is contaminated by food, during the last 20 days before the ship enters port.
5. The 1993 amendments required a report to the Congress by the Navy regarding a plan for compliance with the MARPOL provisions by November 31, 1996. The plan stated, in essence, that the MARPOL provisions were not operationally feasible, but that plastic processors, metal shredders, and glass processors could be installed. The 1996 amendments to APPS allows a discharge, without regard to Special Areas, of a slurry of seawater, paper, cardboard, or food wastes capable of passing through a screen with openings no larger than 12 mm in diameter and greater than 3 nm of land, and metal and glass that have been shredded and bagged so as to ensure negative buoyancy greater than 12 nm from land.

5 Air

5.1 CLEAN AIR ACT

The Clean Water Act served as a model for the construction of the Clean Air Act. There is, as a result, considerable similarity in the methods to control pollution in the two acts as shown in Table 5.1.

1. Early regulations:
   a. Chicago and Cincinnati passed smoke emission controls in 1888.
   b. Pittsburgh and New York followed in the 1890s.
2. Regulation of hazardous air pollutants:

Asbestos	1973
Beryllium	1973
Mercury	1973
Vinyl chloride	1975
Benzene	1977
Radionuclides	1979
Arsenic	1980
Coke oven emissions	1984

TABLE 5.1
Similarities Between the Clean Air Act and Clean Water Act in Methods to Control Pollution

Clean Air Act Requirements	Clean Water Act Requirements
Section 108 Air quality criteria	Section 304 (a) Water quality criteria
Section 109 National ambient air quality standards (NAAQS)	Section 303 Water quality standards (WQS)
Monitored by 4000 SLAMS; 1000 NAMS	Monitoring program via STORET (Storage— retrieval of data)
Section 110 SIPs for NAAQS	Section 305 (b) State water quality reports
Measures to prevent significant deterioration (PSD)	Antidegradation policy of WQS
Control technology guidelines	Effluent guidelines
Reasonably available control technology (RACT)	Best practicable control technology (BPCT)
Best available control technology (BACT)	Best available control technology (BACT)
New source performance standards (NSPS)	Effluent guidelines for new sources
NESHAPS	Effluent guidelines for toxic pollutants
Section 113 Enforcement	Section 309 Enforcement

3. Reports from states indicate that 2,619,991,261 pounds of chemicals were released to the atmosphere in 1988. States recording greater than 100 million lbs were:

Illinois	104,592,707 lbs
Indiana	110,075,627 lbs
Louisiana	133,070,512 lbs
Ohio	136,453,929 lbs
Tennessee	133,697,458 lbs
Texas	169,936,759 lbs
Utah	119,410,265 lbs
Virginia	119,593,757 lbs

4. Chemicals with greater than 100 million lbs emissions to the atmosphere were:

Toluene	263,448,463 lbs
Methanol	215,367,551 lbs
1,1,1-Trichloroethane	161,675,112 lbs
Xylene	140,264,820 lbs
Chlorine	132,566,904 lbs
Methyl ethyl ketone	126,121,008 lbs
Dichloromethane	114,811,826 lbs

5. Industrial categories with greater than 100 million lbs emissions:

Chemical	754,922,471 lbs
Primary metals	232,958,571 lbs
Paper	202,210,446 lbs
Transportation	201,297,144 lbs
Plastics	158,832,600 lbs
Fabricated metals	117,524,318 lbs
Electrical	115,198,789 lbs

6. Between 1978 and 1987, the following decreased by the percentages shown:

Ozone	16 percent
Lead	88 percent
Sulfur dioxide	35 percent
Carbon monoxide	32 percent
Nitrogen dioxide	14 percent
Particulates	23 percent

7. The Clean Air Act has six titles or subchapters, including:
   a. Air pollution prevention and control (Eliminating Nonattainment).
   b. Emission Standards for Moving Sources.
   c. General Requirements.
   d. Acid Deposition Control.
   e. Permits.
   f. Stratospheric Ozone Protection.

The focus of this book will be on Subchapter I, Air Pollution Prevention and Control.

1. Section 107 provides for the designation of air quality control regions, which are geographic areas that share common air quality concerns. There are 242 air quality regions in the United States.
2. Section 109 requires national ambient air quality standards (NAAQS). There are 6 "criteria" pollutants with primary (health) and secondary (welfare) requirements:
   Particulate matter
   Sulfur dioxide
   Carbon monoxide
   Ozone
   Nitrogen dioxide
   Lead
   These are monitored, nationwide, by about 4000 state and local air monitoring stations (SLAMS) and about 1000 national air monitoring stations (NAMS). The NAAQS serve as surrogates for measuring air quality.
3. Section 110 state implementation plans (SIPs) are required to ensure implementation, maintenance, and enforcement of NAAQS.
   a. Air quality control regions (AQCRs) are in attainment or non-attainment for one or more of the NAAQS. Non-attainment is determined by counting the number of exceedences at the monitoring stations. SIPs include source-specific emission limitations to bring an AQCR non-attainment area into attainment.
   b. States are required in the SIP to adopt measures to prevent significant deterioration (PSD) of air quality in clean air areas.
   c. Control technology guidelines (similar to effluent guidelines) are developed to assist states to choose the right controls for existing stationary sources in non-attainment areas.
   d. Reasonably available control technology (RACT) is required for existing sources.
   e. Best available control technology (BACT) economically achievable is required for new sources.
4. Section 111 new source performance standards (NSPS), including allowable emission limitations, are developed and required to be attained by new sources of air pollution.
5. Section 112 national emission standards for hazardous air pollutants (NESHAPs):
   a. There are 189 hazardous air pollutants listed in Act. See Table 5.2.
   b. In Section 112 (b) (4), the Act provides that the Administrator "may use any authority available" to acquire the health or environmental effects of a substance.
   c. The NESHAP for asbestos, 40 CFR 61.140, e.g., addresses:
   Definitions.
   Standard for asbestos mills.
   Standard for manufacturing.
   Standard for demolition and renovation.
   Standard for spraying.

TABLE 5.2
Hazardous Air Pollutants

Acetaldehyde	2,4-D, salts and esters	Formaldehyde
Acetamide	DDE	Heptachlor
Acetonitrile	Diazomethane	Hexachlorobenzene
Acetophenone	Dibenzofurans	Hexachlorobutadiene
2-Acetylaminofluorene	1.2-Dibromo-3-chloropropane	Hexachlorocyclopentadiene
Acrolein	Dibutylphthalate	Hexachloroethane
Acrylamide	1,4-Dichlorobenzene (p)	Hexamethylene-1,6-diisocya-
Acrylic acid	3,3-Dichlorobenzidene	nate
Acrylonitrile	Dichloroethyl ether (bis-2-	Hexamethylphosphoramide
Allyl chloride	chloroethyl-ether)	Hexane
4-Aminobiphenyl	1,3-Dichloropropene	Hydrazine
Aniline	Dichlorvos	Hydrochloric acid
o-Anisidine	Diethanolamine	Hydrogen fluoride
Asbestos	N,N-Diethyl aniline	(hydrofluoric acid)
Benzene	(N,N-dimethylaniline)	Hydrogen sulfide
Benzidine	Diethyl sulfate	Hydroquinone
Benzotrichloride	3,3-Dimethoxybenzidine	Isophorone
Benzyl chloride	Dimethyl aminoazobenzene	Lindane
Biphenyl	3,3'-Dimethyl benzidine	Maleic anhydride
bis-2-Ethylhexyl-phthalate	Dimethyl carbamoyl chloride	Methanol
(DEHP)	Dimethyl formamide	Methoxychlor
bis-Chloromethyl-ether	1,1-Dimethyl hydrazine	Methyl bromide
Bromoform	Dimethyl phthalate	(bromomethane)
1,3-Butadiene	Dimethyl sulfate	Methyl chloride
Calcium cyanamide	4,4-Dinitro-o-cresol and salts	(chloromethane)
Caprolactam	2,4-Dinitrophenol	Methyl chloroform
Captan	2,4-Dinitrotoluene	(1,1,1-tri-chloroethane)
Carbaryl	1,4-Dioxane	Methyl ethyl ketone
Carbon disulfide	(1,4-diethyleneoxide)	(2-butanone)
Carbon tetrachloride	1,2-Diphenylhydrazine	Methyl hydrazine
Carbonyl sulfide	Epichlorohydrin (1-chloro-2,3-	Methyl iodide (iodomethane)
Catechol	epoxypropane)	Methyl isobutyl ketone (hexone)
Chloramben	1,2-Epoxybutane	Methyl isocyanate
Chlordane	Ethyl acrylate	Methyl methacrylate
Chlorine	Ethyl benzene	Methyl tert butyl ether
Chloroacetic acid	Ethyl carbamate (urethane)	4,4-Methylene
2-Chloroacetophenone	Ethyl chloride (chloroethane)	bis-2-chloroaniline
Chlorobenzene	Ethylene dibromide	Methylene chloride
Chlorobenzilate	(dibromomethane)	(dichloromethane)
Chloroform	Ethylene dichloride	Methylene diphenyl
Chloromethyl methyl ether	(1,2-dichloroethane)	diisocyanate
Chloroprene	Ethylene glycol	4,4-Methylenedianiline
Cresols/cresylic acid	Ethylene imine (aziridine)	Naphthalene
o-Cresol	Ethylene oxide	Nitrobenzene
m-Cresol	Ethylene thiourea	4-Nitrobiphenyl
p-Cresol	Ethylidene dichloride	4-Nitrophenol
Cumene	(1,1-dichloroethane)	2-Nitropropane

N-Nitroso-N-methylurea	Styrene	Vinylidene chloride
N-Nitrosodimethylamine	Styrene oxide	(1,1-dichloroethylene)
N-Nitrosomorpholine	2,3,7,8-Tetrachlorodibenzo-	Xylenes
Parathion	p-dioxin	o-Xylenes
Pentachloronitrobenzene	1,1,2,2-Tetrachloroethane	m-Xylenes
(quintobenzene)	Tetrachloroethylene	p-Xylenes
Pentachlorophenol	(perchloroethylene)	Antimony compounds
Phenol	Titanium tetrachloride	Arsenic compounds
p-Phenylenediamine	Toluene	Beryllium compounds
Phosgene	2,4-Toluene diamine	Cadmium compounds
Phosphine	2,4-Toluene diisocyanate	Chromium compounds
Phosphorus	o-Toluidine	Cobalt compounds
Phthalic anhydride	Toxaphene (chlorinated	Coke oven compounds
Polychlorinated biphenyls	camphene)	Cyanide compounds
1,3-Propane sultone	1,2,4-Trichlorobenzene	Glycol compounds
β-Propiolactone	1,1,2-Trichloroethane	Lead compounds
Propionaldehyde	Trichloroethylene	Manganese compounds
Propoxur (baygon)	2,4,5-Trichlorophenol	Mercury compounds
Propylene dichloride	2,4,6-Trichlorophenol	Fine mineral fibers
(1,2-dichloropropane)	Triethylamine	Nickel compounds
Propylene oxide	Trifluralin	Polycyclic organic matter
1,2-Propylenimine	2,2,4-Trimethylpentane	Radionuclides (including
(2-methyl aziridine)	Vinyl acetate	radon)
Quinoline	Vinyl bromide	Selenium compounds
Quinone	Vinyl chloride	

Standard for fabricating.

Standard for insulating materials.

Standard for waste disposal for asbestos mills.

Standard for waste disposal for manufacturing, fabricating, demolition, renovation, and spraying operations.

Standard for inactive waste disposal sites for asbestos mills and manufacturing and fabricating operations.

Standard for active waste disposal sites.

5.2   ASBESTOS

1. Prior to mid-1970s, asbestos was widely used as insulation because of its high tensile strength, good heat and electrical insulating properties, and good chemical resistance. It also was used in floor tile, ceiling tile, brake shoes, corrugated-like paper products used for thermal system insulation, gaskets in heating and air conditioning equipment, cement asbestos water pipe, fire doors, fire brick for boilers, spray-applied or troweled-on materials on walls and ceilings, transite wall board, and other uses.

2. Asbestos occurs naturally as a mineral in six forms:

   Chrysotile   white
   Amosite   brown

      Crocidolite   blue
      Tremolite
      Anthophyllite
      Actinolite

    Asbestos is composed of SiO_2, MgO, FeO, Fe_2O_3, and H_2O.

3. When processed, asbestos breaks down into small diameter fibers that can be inhaled and the fibers readily pass through lung tissue. Following repeated or excessive exposure for 20 to 30 years, health problems can arise such as:

      Asbestosis   fibrosis of the lung
      Mesothelioma   cancer of the chest lining
      Lung cancer
      GI tract cancers   larynx, stomach, colon, and rectum

4. OSHA and EPA are the principal agencies regulating asbestos:

   a. In late 1994, OSHA lowered the permissible exposure level (PEL) for asbestos to 0.1 fiber per cc of air over an 8 h work day. If exposed to the PEL, workers must have:

      Daily personal air monitoring.
      Employee notification.
      Respiratory protection program.
      Medical surveillance.
      Physical examination, including chest X-ray and pulmonary function test.

   b. The Asbestos Hazard Emergency Response Act (AHERA) was signed into law in October 1986 as Title 2 of the Toxic Substances Control Act (TSCA). The law requires the EPA to develop regulations which provide a comprehensive framework for addressing asbestos problems in public and private elementary and secondary schools.

5. Asbestos inspectors, workers, supervisors, and contractors who remove or repair asbestos must be licensed in many states. Accreditation entails a one- to two-week intensive training initially and a one-day refresher course annually thereafter.

6. Concerns:

   a. Any material that contains 1 percent or more asbestos is asbestos containing material (ACM).

   b. Principal concern is with friable asbestos. Friable asbestos is asbestos that can be crushed between the thumb and forefingers.

   c. Secondary concern is with non-friable material that may become friable. Non-friable material may become friable (potential for fibers to enter atmosphere) because of:

      Sharp object contact with surface.
      Vibration.
      Air erosion from air ducts.
      Water stains.

   d. Options for asbestos control:

      Operation and maintenance (surveillance every six months).

Repair.

Encapsulation.

Enclosure.

Removal.

7. To sample asbestos surfacing materials:

Group areas into homogenous areas (uniform in texture and appearance; installed at same time; same general use areas).

Diagram and divide into nine equally sized subareas.

For area greater than 5000 ft2, take seven samples.

For area greater than 1000, but less than 5000 ft2, take five samples.

For area less than 1000 ft2, take three samples.

For each homogenous pipe area, take three samples.

8. No regulated asbestos containing material may be removed unless an on-site management person is trained in the asbestos regulation (NESHAP).

9. Always keep asbestos containing materials wet when working with them in any manner.

10. NESHAP standards for waste disposal:

a. Discharge no visible emissions to the outside air during collection, processing, packaging, or transportation.

b. After wetting, seal all asbestos containing materials in leak-tight containers.

c. Label containers with warning labels, name of waste generator, and location location where generated.

5.3   OTHER SECTIONS OF THE CLEAN AIR ACT

Section 309 Provides for EPA review and comment in writing on the environmental impact of any matter related to any major federal agency action, legislation, or newly authorized federal project. Such review and comments shall be made available to the public. In the event the EPA determines that an action is unsatisfactory, the EPA shall publish the determination and the matter shall be referred to the President's Council on Environmental Quality (CEQ). Section 309 is an important section of law that is not restricted to clean air activities. It is principally associated with the National Environmental Policy Act and ensures that the EPA provide comments on environmental assessment activities of other federal agencies and that those comments are made public.

Section 328 controls air pollution from outer continental shelf activities to a distance of 25 nm beyond a state's seaward boundary.

5.4   AIR GLOSSARY

The U.S. Environmental Protection Agency has published a glossary of air related terms that is worthy of presenting here:

Acid rain—Air pollution produced when acid chemicals are incorporated into rain, snow, fog, or mist. The "acid" in acid rain comes from sulfur oxides and nitrogen oxides, products of burning coal and other fuels and from certain industrial processes. The sulfur oxides and nitrogen oxides are related to two strong acids: sulfuric acid and nitric acid. When sulfur dioxide and nitrogen oxides are released from power plants and other sources, winds blow them far from their source. If the acid chemicals in the air are blown into areas where the weather is wet, the acids can fall to earth in the rain, snow, fog, or mist. In areas where the weather is dry, the acid chemicals may become incorporated into dusts or smokes. Acid rain can damage the environment, human health, and property.

Attainment area—A geographic area in which levels of a criteria air pollutant meet the health-based primary standard (national ambient air quality standard or NAAQS) for the pollutant. An area may have an acceptable level for one criteria air pollutant, but may have unacceptable levels for others. Thus, an area could be both attainment and non-attainment at the same time. Attainment areas are defined using federal pollutant limits set by EPA.

Carbon monoxide (CO)—A colorless, odorless, poisonous gas, produced by incomplete burning of carbon-based fuel including gasoline, oil, and wood. Carbon monoxide is also produced from incomplete combustion of many natural and synthetic products. For instance, cigarette smoke contains carbon monoxide. When carbon monoxide gets into the body, the carbon monoxide combines with chemicals in the blood and prevents the blood from bringing oxygen to cells, tissues, and organs. The body's parts need oxygen for energy, so high-level exposures to carbon monoxide can cause serious health effects, with death possible from massive exposures. Symptoms of exposure to carbon monoxide can include vision problems, reduced alertness, and general reduction in mental and physical functions. Carbon monoxide exposures are especially harmful to people with heart, lung, and circulatory system diseases.

Chlorofluorocarbons (CFCs)—These chemicals and some related chemicals have been used in great quantities in industry, for refrigeration and air conditioning, and in consumer products. CFCs and their relatives, when released into the air, rise into the stratosphere, a layer of the atmosphere high above the Earth. In the stratosphere, CFCs and their relatives take part in chemical reactions which result in reduction of the stratospheric ozone layer, which protects the Earth's surface from harmful effects of radiation from the sun. The 1990 Clean Air Act includes provisions for reducing releases (emissions) and eliminating production and use of these ozone-destroying chemicals.

Combustion—Burning; many important pollutants, such as sulfur dioxide, nitrogen oxides, and particulates are combustion products, often products of the burning of fuels such as coal, oil, gas, and wood.

Continuous emission monitoring systems (CEMS)—Machines which measure, on a continuous basis, pollutants released by a source. The 1990 Clean Air Act requires continuous emission monitoring systems for certain large sources.

Control technology; control measures—Equipment, processes, or actions used to reduce air pollution. The extent of pollution reduction varies among technologies and measures. In general, control technologies and measures that do the best job of reducing pollution will be required in the areas with the worst pollution. For example, the best available control technology or best available control measures (BACT, or BACM) will be required in serious non-attainment areas for particulates, a criteria air pollutant. A similar high level of pollution reduction will be achieved with maximum achievable control technology (MACT) which will be required for sources releasing hazardous air pollutants.

Emission—Release of pollutants into the air from a source.

Enforcement—The legal methods used to make polluters obey the Clean Air Act. Enforcement methods include citations of polluters for violations of the law (citations are much like traffic tickets), fines, and even jail terms. The EPA and state and local governments are responsible for enforcement of the Clean Air Act, but if they do not enforce the law, members of the public can sue the EPA or the states to get action. Citizens can also sue violating sources, apart from any action the EPA or state or local governments have taken. In some cases the EPA can fine violators without going to court first. The purpose of this new authority is to speed up violating sources' compliance with the law and reduce court time and cost.

Hazardous air pollutants (HAPs)—Chemicals that cause serious health and environmental effects. Health effects include cancer, birth defects, nervous system problems, and death owing to massive accidental releases such as occurred at the pesticide plant in Bhopal, India. Hazardous air pollutants are released by sources such as chemical plants, dry cleaners, printing plants, and motor vehicles.

Material safety data sheets (MSDS)—Product safety information sheets prepared by manufacturers and marketers of products containing toxic chemicals; these sheets can be obtained by requesting them from the manufacturer or marketer.

Nitrogen oxides (NO_x)—A criteria air pollutant; nitrogen oxides are produced from burning fuels, including gasoline and coal. Nitrogen oxides are smog formers and react with volatile organic compounds to form smog. Nitrogen oxides are also major components of acid rain.

Non-attainment areas—A geographic area in which the level of a criteria air pollutant is higher than the level allowed by the federal standards. A single geographic area may have acceptable levels of one criteria air pollutant, but unacceptable levels of one or more other criteria air pollutants. Thus, an area can be both attainment and non-attainment at the same time. It has been estimated that 60 percent of Americans live in non-attainment areas.

Offset—A method used in the 1990 Clean Air Act to give companies which own or operate large sources in non-attainment areas flexibility in meeting overall pollution reduction requirements when changing production processes. If the owner or operator of the source wishes to increase releases of a criteria air pollutant, an offset (reduction of a somewhat greater amount of the same pollutant) must be obtained either at the same plant or by purchasing offsets from another company.

Oxygenated fuel (oxyfuel)—Special type of gasoline, which burns more completely than regular gasoline in cold start conditions; more complete burning results in reduced production of carbon monoxide, a criteria air pollutant. In some parts of the country, carbon monoxide releases from cars starting up in cold weather makes a major contribution to pollution. In these areas, gasoline refiners must market oxygenated fuels, which contain a higher oxygen content than regular gasoline.

Ozone—A gas which is a variety of oxygen. The oxygen gas found in the air consists of two oxygen atoms stuck together; this is molecular oxygen. Ozone consists of three oxygen atoms stuck together into an ozone molecule. Ozone occurs in nature; it produces the sharp smell noticed near a lighting strike. High concentrations of ozone gas are found in a layer of the atmosphere—the stratosphere—high above the Earth. Stratospheric ozone shields the Earth against harmful rays from the sun, particularly ultraviolet B. Smog's main component is ozone; this ground-level ozone is a product of reactions among chemicals produced by burning coal, gasoline, and other fuels, and chemicals found in products including solvents, paints, and hairsprays.

Ozone hole—Thin place in the ozone layer located in the stratosphere high above the Earth. Stratospheric ozone thinning has been linked to destruction of stratospheric ozone by CFCs and related chemicals. Ozone holes have been found above Antarctica and above Canada and northern parts of the United States, as well as above northern Europe.

Particulates; particulate matter—Particulate matter includes dust, soot and other tiny bits of solid materials that are released into and move around in the air. Particulates are produced by many sources, including burning of diesel fuels by trucks and buses, incineration of garbage, mixing and application of fertilizers and pesticides, road construction, industrial processes such as steelmaking, mining operations, agricultural burning and operation of fireplaces and woodstoves. Particulate pollution can cause eye, nose, and throat irritation, and other health problems.

Permit—A document that resembles a license, required by the Clean Air Act for major sources of air pollution, such as power plants, chemical factories and, in some cases, smaller polluters. Usually permits will be given out by states, but if the EPA has disapproved part or all of a state permit program, the EPA will give out the permits in that state. The 1990 Clean Air Act includes requirements for permit applications, including provisions for members of the public to participate in state and EPA review of permit applications. Permits include information on which pollutants are being released, how much the source is allowed to release, and the program that will be used to meet pollutant release requirements. Permits are required both for the operation of plants and for the construction of new plants.

State implementation plan (SIP)—A detailed description of the programs a state will use to carry out its responsibilities under the Clean Air Act. State implementation plans are collections of the regulations used by a state to reduce air pollution. The Clean Air Act requires that the EPA approve each state implementation plan. Members of the public are given opportunities to participate in review and approval of state implementation plans.

Sulfur dioxide—A criteria air pollutant; sulfur dioxide is a gas produced by burning coal, most notably in power plants. Some industrial processes, such as production of paper and smelting of metals produce sulfur dioxide. Sulfur dioxide is closely related to sulfuric acid, a strong acid. Sulfur dioxide has an important role in the production of acid rain.

Temperature inversion—A weather condition that is often associated with serious smog. In a temperature inversion, warm air does not rise because it is trapped near the ground by a layer of heavy colder air above it. Pollutants in the warm air, especially smog and smog-forming chemicals, including volatile organic compounds, are trapped close to the ground. As people continue driving, and sources other than motor vehicles continue to release smog-forming pollutants into the air, the smog level keeps getting worse.

Ultraviolet B (UVB)—A type of sunlight; the ozone in the stratosphere, high above the Earth, filters out ultraviolet B rays and keeps them from reaching the Earth. Ultraviolet B exposure has been associated with skin cancer, eye cataracts, and damage to the environment. Thinning of the ozone layer in the stratosphere results in increased amounts of ultraviolet B reaching the Earth.

Volatile organic compounds (VOCs)—Organic chemicals all contain the element carbon; organic chemicals are the basic chemicals found in living things and in products derived from living things, such as coal, petroleum, and refined petroleum products. Many of the organic chemicals we use do not occur in nature, but were synthesized by chemists in laboratories. Volatile chemicals produce vapors readily at room temperature and normal atmospheric pressure. Vapors escape easily from volatile liquid chemicals. Volatile organic chemicals include gasoline, industrial chemicals such as benzene, solvents such as toluene and xylene, and tetrachloroethylene, the principal dry cleaning solvent. Many volatile organic chemicals are also hazardous air pollutants; e.g., benzene causes cancer.

6 Land

6.1 RESOURCE CONSERVATION AND RECOVERY ACT (RCRA)

The Resource Conservation and Recovery Act (RCRA) of 1976 was enacted for two principal purposes: to regulate hazardous wastes from their generation through their disposal, and to protect the groundwater aquifers from the land disposal of hazardous wastes.

1. The Hazardous and Solid Waste Amendments of 1984 to RCRA were noteworthy; they:
   a. Prohibited bulk non-containerized liquid hazardous waste in any landfill.
   b. Required that disposal of containerized liquid hazardous wastes in landfills be minimized.
   c. Regulated small quantity generators, i.e., 100 to 1000 kg/mo.
   d. Regulated underground storage tanks.
2. Section 3001 Identification and Listing of Hazardous Waste:
   a. Requires criteria for identifying and listing of hazardous waste, "taking into account toxicity, persistence, and degradability in nature, potential for accumulation in tissue, and other related factors such as flammability, corrosiveness, and other hazardous characteristics."
   b. A solid waste (discarded or intended to be discarded liquid, semisolid, contained gaseous material, sludge or refuse) is a hazardous waste if it:
      - Exhibits the characteristics of a hazardous waste:
        Ignitability with flash point less than 140°F, 40 CFR 261.21.
        Corrosivity with pH less than or equal to 2 or greater than or equal to 12.5, 40 CFR 261.22.
        Reactivity by reacting violently with water, 40 CFR 261.23.
        Toxicity where toxicity characteristics leaching procedure (TCLP) leachate greater than or equal to 100 times the primary drinking water criteria concentrations, 40 CFR 261.24.
      - Listed as waste from a non-specific source (e.g., F006 wastewater treatment sludges from electroplating operations) 40 CFR 261.31; or listed as waste from a specific source (e.g., wastewater treatment sludges from the manufacturing and processing of explosives) 40 CFR 261.32.
      - Is a discarded commercial chemical product, 40 CFR 261.33.
      - By applying the generators personal knowledge of the hazard characteristics of the waste, 40 CFR 262.11. See Table 6.1.

49

TABLE 6.1
Toxicity Characteristics Leaching Procedure Maximum Concentration of Contaminants

	Regulatory Level (mg/1)		Regulatory Level (mg/1)
Arsenic	5.0	Hexachlorobutadiene	0.15
Barium	100.0	Hexachloroethane	3.0
Benzene	0.5	Lead	5.0
Cadmium	1.0	Lindane	0.4
Carbon tetrachloride	0.5	Mercury	0.2
Chlordane	0.03	Methoxychlor	10.0
Chlorobenzene	100.0	Methyl ethyl ketone	200.0
Chloroform	6.0	Nitrobenzene	2.0
Chromium	5.0	Pentachlorophenol	100.0
Cresol	200.0	Pyridine	5.0
2,4-D	10.	Selenium	1.0
1,4-Dichlorobenzene	7.5	Silver	5.0
1,2-Dichloroethane	0.5	Tetrachloroethylene	0.7
1,1-Dichloroethylene	0.7	Toxaphene	0.5
2,4-Dinitrotoluene	0.13	Trichloroethylene	0.5
Endrin	0.02	2,4,5-Trichlorophenol	400.0
Heptachlor	0.008	2,4,6-Trichlorophenol	2.0
(and its epoxide)		2,4,5-TP Silvex	1.0
Hexachlorobenzene	0.13	Vinyl chloride	0.2

3. Section 3002 Standards Applicable to Generators of Hazardous Waste:
   a. Requires regulations establishing standards applicable to generators of hazardous wastes, "as may be necessary to protect human health and the environment."
   b. EPA identification number required, 40 CFR 262.12.
   c. Manifest required for transporting waste off-site, 40 CFR 262.20.
      • Generator must retain records for three years.
      • Must file exceptions report to the EPA if signed manifest copy is not received from treatment, storage, or disposal facility operator within 45 days, 40 CFR 262.42.
   d. On-site hazardous waste storage time, without a Part B permit, shall not exceed 90 days, except a small quantity generator (greater than 100 and less than 1000 kg/mo) may store hazardous waste for 180 days.
   e. Small quantity generators need only ensure proper waste disposal and comply with certain reporting requirements.
4. Section 3003 Standards Applicable to Transporters:
   a. EPA identification number required, 40 CFR 263.11.
   b. Must comply with manifest system and retain copies of records for three years, 40 CFR 263.20.

    c. Must adhere to spill reporting and cleanup procedures in event of spill, 40 CFR 263.30.

5. Section 3004 Standards Applicable to Hazardous Waste Treatment, Storage, and Disposal (TSD) Facilities:

    a. Requires that regulations be promulgated for:
  - Minimizing disposal of containerized liquid hazardous waste in landfills.
  - Prohibiting disposal of non-containerized liquids in landfills.

    b. Regulations governing treatment, storage, and disposal facilities, 40 CFR Part 264, require:
  - EPA identification number.
  - Waste analysis as received.
  - Implementation of security plan (Keep Out signs).
  - Inspection of monitoring equipment.
  - Training plan and personnel training.
  - Development of contingency plan with copy to police, fire, and hospital facilities.
  - Developing of emergency procedures and designation of emergency coordinator.
  - Compliance with manifest system.
  - Implementation of groundwater monitoring program.
  - Development of written closure plan.
  - Post closure plan.
  - Financial assurance for closure.
  - Financial assurance for post closure.

6. Regulatory Permits for TSD Facilities:

    a. Part A permit contents:
  - Facility name and mailing address.
  - Operator name, address, and telephone number.
  - Description of processes to be used at TSD.
  - Listing of permits obtained and applied for.
  - Topographic map of facility.

    b. Part B permit contents:
  - Description of facility.
  - Chemical and physical analyses of hazardous waste to be handled at facility.
  - Waste analysis plan.
  - Description of security procedures.
  - Inspection schedule.
  - Contingency plan.
  - Description of procedures to prevent.
    Spilling during unloading.
    Runoff.
    Contamination of water supplies.
    Undue exposure of personnel.

- Precautions to prevent ignition or reactivity of incompatible wastes.
- Traffic pattern plan.
- Facility location information.
- Flood plain information.
- Training program plan.
- Closure plan.
- Closure cost estimate.
- Insurance policy.
- Topographic map.
- Groundwater monitoring program.
- Groundwater monitoring data.
- Specific information required related to containers, tanks, waste piles, surface impoundments, incinerators, land treatment facilities, and landfills.

7. Section 3007 Inspections:
   a. EPA/State shall inspect TSD facilities of federal agencies annually.
   b. Agencies shall reimburse the EPA for costs of the inspection of the facility.
8. Section 3008 Federal Enforcement:
   a. Any person who knowingly transports, treats, stores, disposes of, or exports any hazardous waste or used oil who knows at that time that he thereby places another person in imminent danger of death or serious bodily injury, shall, upon conviction, be subject to a fine of not more than $250,000 or imprisonment for not more than 15 years, or both. A defendant that is an organization shall, upon conviction of violation, be subject to a fine of not more than $1 million.
9. Section 3022 Public Vessels:
   a. Any hazardous waste generated on a public vessel shall not be subject to the storage, manifest, inspection, or record-keeping requirements of RCRA until such waste is transferred to a shore facility, unless the waste is stored on the public vessel for more than 90 days after it is placed in reserve or otherwise no longer in service, or the waste is transferred to another public vessel within the territorial waters and is stored for more than 90 days after the date of transfer.
10. Section 6001 Federal Facilities:
    a. The United States expressly waives any immunity with respect to substantive or procedural requirements, including injunctive relief, administrative order, or civil or administrative penalty or fine, or reasonable service charge. The service charges include fees or charges assessed in connection with processing and issuing permits, renewal of permits, plan review, and inspection and monitoring of facilities in connection with a federal, state, interstate, or local solid waste or hazardous waste regulatory program.
    b. No agent, employee, or officer of the United States shall be personally liable for any civil penalty under any federal, state, interstate, or local solid or hazardous waste law, however, they shall be subject to any criminal sanctions under such laws.

11. Underground Storage Tanks (USTs):
    a. Section 9001 defines a UST as any tank with greater than 10 percent of its volume buried below ground, including attached pipes. Farm or residential tanks of 1,100 gallons or less used for storing motor fuel for non-commercial purposes and tanks used for storing heating oil for consumptive use on the premises are not included.
    b. Section 9003 requires regulations for:
       • Tank testing and maintaining a leak detection system.
       • Maintaining records of release detection methods and releases.
       • Report releases and corrective actions.
       • Corrective action in response to a release.
       • Tank construction to control spill and overfill, prevent releases owing to corrosion or structural failure, with cathodical protection against corrosion if required and lining that is compatible with the substance to be stored.
       • Provide for proper tank closure.
       • Provide evidence, if required, of financial capability to complete the above.
12. Report to Congress on Injection of Hazardous Waste—the report shall include:
    a. The location and depth of each well.
    b. Engineering and construction details of each well.
    c. Hydrogeological characteristics of overlying and underlying strata.
    d. Location and size of all drinking water aquifers penetrated by the well.
    e. Location, capacity, and population served by each well providing drinking or irrigation water within a 5-mi radius of the injection well.
    f. Nature and volume of the waste injected during the past year.
    g. Dates and nature of inspections of the injection well.
    h. Name and address of all owners and operators of the well and any disposal facility associated with it.
    i. Records of any enforcement action and an identification of the wastes involved in such enforcement action.

6.2 THE COMPREHENSIVE ENVIRONMENTAL RESPONSE, COMPENSATION, AND LIABILITY ACT OF 1980 (CERCLA; SUPERFUND)

6.2.1 A SIMPLIFIED CONCEPT

1. Abandoned dumps that create Love Canal situations, harming health and the environment, need to be cleaned up.
2. Levying a tax on the manufacture of certain chemicals creates a trust fund called the Superfund.
3. Sites to be cleaned up are prioritized by the EPA in a National Priority List (NPL).
4. Cleanup criteria and procedures are specified in a National Contingency Plan (NCP).

5. Government was given authority to use funds from Superfund to cleanup sites on NPL using NCP procedures. Remedial investigation/feasibility studies (RI/FS) became the first significant step in the process.
6. Cleanup costs were to be recovered through the court system from those responsible for creating the waste site through the law's liability provisions, with the recovered money going back into the Superfund and providing a continuing revolving fund for abandoned waste dump cleanup activities.

6.2.2 PRINCIPAL PROVISIONS

1. CERCLA was enacted to cleanup abandoned and inactive hazardous waste sites.
2. Section 104 provides that the EPA can act, consistent with the National Contingency Plan (NCP), to remove or arrange for the removal of hazardous wastes and provide for remedial action, when there is a release, or substantial threat of a release, which may present an imminent and substantial danger to the public health or welfare.
3. Section 105 requires revision and update of the NCP for the removal of oil and hazardous substances, originally prepared and published pursuant to Section 311 of the Clean Water Act, and publication of a priority list of sites to be cleaned up. The NCP is codified at 40 CFR Part 300.
   a. The revisions to the NCP must include:
      • Methods for discovering and investigating facilities at which hazardous substances have been disposed.
      • Methods for evaluating, including analyses of relative cost, and remedying any releases or threats of releases that post substantial danger to the public health or the environment.
      • Methods and criteria for determining the appropriate extent of removal, remedy, and other measures.
      • Appropriate roles and responsibilities for the federal, state, and local governments.
      • Provision for identification, procurement, maintenance, and storage of response equipment and supplies.
      • A method for and assignment of responsibility for reporting the existence of facilities on federally-owned or controlled properties and any releases of hazardous substances from these facilities.
      • Means of assuring that remedial action measures are cost effective over the period of potential exposure to the hazardous substances.
      • Criteria for determining priorities among releases or threatened releases for purposes of taking remedial action.
      • Specified roles for private organizations and entities in preparation for response and in responding to releases of hazardous substances.
   b. Section 105(a)(8)(B) of CERCLA requires the EPA to promulgate a National Priority List (NPL) to list sites with known or threatened releases of hazardous substances, pollutants, or contaminants. The NPL

is found at Appendix B of 40 CFR 300. There are three mechanisms for placing a site on the NPL:

- The site scores sufficiently high on the hazardous ranking system (HRS). This system provides numerical scores for assessed harm to humans, or the environment, from migration of hazardous substances by groundwater, surface water, or air; potential harm from substances that can explode or cause fires, or potential harm from direct contact with hazardous substances.
- Each state may designate a single site as its top priority, regardless of the HRS score.
- Sites may be listed regardless of their HRS scores based upon health and threat potential or an EPA decision that it will be more cost effective to use remedial (structured evaluation) authority rather than (immediate) removal authority.

  c. EPA promulgated an original NPL of 406 sites (48 FR 40658, September 8, 83). The NPL now contains about 1,200 sites in the General Superfund Section and 150 in the Federal Facilities Section (60 FR 20330, April 25, 1995). There are about 30 thousand potential hazardous waste sites on state lists being evaluated for potential proposal to the NPL.

  d. The EPA may delete sites from the NPL where no further response is considered appropriate under Superfund.

  e. Remedial action at a site may be financed by the Superfund Trust Fund *only* after it is placed on the NPL.

  f. Placing a site on the NPL does not guarantee that federal funds will be expended. Liability clauses of CERCLA could be invoked to reclaim trust funds spent.

4. The federal response process includes:

  a. Site discovery or notification may take place through reports submitted in regulatory compliance, investigations by government authorities, inventory or survey efforts, or random or incidental observations. The EPA is authorized to begin response actions immediately using trust fund monies. Responsible parties can undertake a response action as a result of EPA enforcement. States can act using state trust funds pursuant to the agreement with the EPA.

  b. Removal actions may occur at preliminary assessment or at any time thereafter. Removal begins with a preliminary evaluation to determine the identity of the source and nature of the release or threat of release, an evaluation of the threat to public health and the magnitude of the threat, and an evaluation of factors necessary to make the determination of whether a removal is necessary. When it is found that fire, explosion, contamination of drinking water, or other hazardous releases are potentials, removal action may be instituted.

  c. During site inspection, inspectors visit the site, assess the potential for immediate pollution of the environment, and collect samples.

d. The EPA has developed a hazard ranking system to govern the placement of a site on the National Priority List.

e. Enforcement is a significant effort on the parts of the EPA and the states to identify potentially responsible parties and to compel them to undertake the required cleanup activities through legal action if necessary. If this cannot be done, the EPA or the respective state will proceed with the cleanup using trust fund resources, and will attempt to recover the costs later.

f. The RI/FS purpose is to characterize the nature and extent of contamination, the likely exposure pathways at a site, the extent of risk raised by the contamination, and the selection and evaluation of potential remedies to mitigate the extent of damage. The development of an RI/FS includes project scoping, data-collection risk assessment, and analysis of alternatives. The no-action alternative must be developed.

g. The record of decision (ROD) documents the selection of a remedy. It contains all facts, analyses of facts, and site-specific policy determinations considered in the course of carrying out activities related to the response effort.

h. Remedial design includes the preparation of detailed engineering plans, drawings, and specifications to implement the chosen remedial alternative.

i. Nine criteria are required to be evaluated in the remedial action process (40 CFR 300.430).
   • Overall protection of human health and the environment.
   • Compliance with ARARs; alternatives must be assessed to determine whether they attain applicable or relevant and appropriate requirements of other federal and state environmental public health laws and regulations.
   • Long-term effectiveness and permanence, i.e., the degree of certainty that the remedial alternative will prove successful.
   • Reduction of toxicity, mobility, or volume of hazardous waste.
   • Short-term effectiveness, including short-term risks to the local community.
   • Implementability, i.e., the ease or difficulty of implementing the response action.
   • Costs.
   • State acceptance.
   • Community acceptance.

5. Section 121 provides cleanup standards and applicable, relevant, and appropriate requirements (ARARs).

6. Title II of the Act provides for the establishment of a Trust Fund through imposing a tax on certain chemicals sold by the manufacturer, producer, or importer, as well as a leaking underground storage tank (LUST) Trust Fund through imposing a tax on motor fuels and on fuel used in commercial transportation on inland waterways.

7. The EPA calculates the average cost of cleanup (60 FR 20330, April 25, 1995) per site as:

RI/FS	$1,350,000
Remedial design	1,260,000
Remedial action	22,500,000
Operation and maintenance	5,630,000
Total	$30,740,000

7 Toxic and Hazardous Materials and Wastes

7.1 TOXIC SUBSTANCES CONTROL ACT OF 1976 (TSCA)

The toxic Substances Control Act is based on the concept that any chemical impacting any media that may present an *unreasonable risk to health or the environment* must be controlled.

1. Amended in 1986 by Asbestos Hazard Emergency Response Act (AHERA).
2. Findings by Congress:
   a. Some chemical substances and mixtures may present an unreasonable risk to health or the environment.
   b. Effective regulation of interstate commerce also necessitates regulation of intrastate commerce.
   c. More than 60,000 chemical substances are presently manufactured or processed for commercial use, and 1,000 more are introduced each year.
3. Policy:
   a. Adequate data, developed by the manufacturers, should be developed on effects of chemical substances on health and the environment.
   b. Adequate authority should exist to regulate substances that present an unreasonable injury to health or the environment.
   c. Authority should be exercised with caution.
4. Section 4, Testing of Chemical Substances and Mixtures:
   a. Authorizes the EPA to require product testing of any substance that may present an unreasonable risk of injury to health or the environment.
   b. Testing may be required by rule to develop data with respect to health or the environment where there is an insufficiency of data and experience to determine the unreasonable risk criterion.
   c. An exemption may be granted when equivalent data are on hand and testing would be duplicative, however, fair and equitable reimbursement of the supplier of the initial testing data is required.
5. Section 5, Manufacturing and Processing Notices:
   a. Manufacturers of a new chemical substance or of a significant new use of an existing chemical are required to provide the EPA with a premanufacture notification (PMN) 90 days prior to manufacture or importation. The PMN must include the identify of the chemical, its molecular

structure, proposed use, and estimate of amount to be manufactured, by-products, exposure estimates, and test data related to human health and environmental effects.

b. If the EPA does not make a declaration to restrict the product, full marketing can begin, and the chemical is added to the existing chemicals inventory.

c. If the EPA determines that the product may present an unreasonable risk to human health or the environment, the EPA may issue an order to limit or prohibit manufacturer, import, processing, distribution in commerce, use, or disposal of the substance, pending development of test data needed to evaluate the potential hazard.

6. Section 6 Regulation of Hazardous Chemical Substances and Mixtures:

a. The EPA may ban, prohibit, or restrict the manufacture, processing, distribution in commerce, or use of chemicals or chemical substances when there is reason to believe that the manufacture, processing, distributing, use, or disposal of a chemical substance may cause an unreasonable risk or injury to health or the environment.

b. Actions that may be taken include:
   • To prohibit the manufacturing, processing, or distribution.
   • To limit the amount that may be manufactured, processed, or distributed.
   • To regulate the concentration for a particular use.
   • To require specific markings or warnings.
   • To require specific disposal requirements.
   • To provide that such actions be limited in application to specified geographic areas.

c. Polychlorinated biphenyls (PCBs) are specifically addressed.
   • The EPA is required to prescribe methods for disposal.
   • PCBs are required to be marked with adequate warning.
   • Manufacturing, processing, distribution in commerce, or use prohibited except in a totally enclosed manner.
   • The EPA may authorize manufacturing, processing, distribution in commerce, or use with a finding that such will not present an unreasonable risk of injury to health or the environment.

7. Section 7, Imminent Hazards:

a. In the event of an imminent and unreasonable risk of serious or widespread injury to health or the environment, civil action in a U.S. District Court is authorized for seizure; notice of risk to the affected parties; or recall, replacement, or repurchase of a substance.

b. The imminent hazard provision is triggered if the hazard is likely to result before a final rule under Section 6 can protect against such risk.

8. The EPA is required to compile, keep current, and publish a list of each chemical substance that is manufactured or processed in the United States.

9. Pesticides, tobacco, nuclear materials, firearms and ammunition, food additives, drugs, and cosmetics are excluded from action under TSCA.

8 Other Environmental Laws

8.1 EXECUTIVE ORDER 12856

Executive Order 12856 requires federal agency compliance with the Emergency Planning and Community Right-to-Know Act (EPCRA) of 1986 and the Pollution Prevention Act (PPA) of 1990 and establishes federal agency goals and compliance date requirements. Initially, federal agencies were excluded from EPCRA because of the Standard Industrial Code applicability clause. However, industrial compliance was required when the Act became effective. The 1984 release of methyl isocyanate from an industrial plant in Bhopal, India, which killed a large number of people, and a later similar release of the same chemical at Institute, WV, in which no one was killed, led to the passage of EPCRA.

1. For each agency facility, the use of 10,000 or more lbs per year of a listed toxic chemical requires compliance and the filing of reports.
2. In addition to the use reporting threshold criterion, industry may meet the reporting threshold through other specified criteria.

8.2 EMERGENCY PLANNING AND COMMUNITY RIGHT-TO-KNOW ACT OF 1986

1. EPCRA requires the governor of each state to designate a State Emergency Response Commission (SERC). Many SERCs include public agencies and departments concerned with issues relating to environment, natural resources, emergency services, public health, occupational safety, and transportation. All governors have established SERCs.
2. The SERC must designate local emergency planning districts and appointed Local Emergency Planning Committees (LEPC) for each district. SERCs have designated over 4000 local districts. Countries are chosen by 35 State Commissions as the basic district designation (often with separate districts for municipalities) and 10 SERCs designated substate planning districts. The SERC is responsible for supervising and coordinating the activities of the LEPC.
3. The LEPC must include, as a minimum, elected state and local officials, police, fire, civil defense, public health professionals, environmental, hospital, and transportation officials, as well as representatives of facilities subject to the emergency planning requirements, community groups, and the media.

4. The LEPC must develop an emergency response plan and review it at least annually thereafter. The plan must identify facilities and transportation routes of extremely hazardous substances; describe emergency response procedures; designate a community coordinator and facility coordinators to implement the plan; outline emergency notification procedures; describe methods for determining the occurrence of a release; describe community and industry emergency equipment and facilities; outline evacuation plans; describe a training program for emergency response personnel; and present methods and schedules for exercising emergency response plans.

8.3   POLLUTION PREVENTION ACT OF 1990

1. The PPA requires collection of mandatory information on source reduction, recycling, and treatment beginning with the 1991 reporting year.
2. Reporting requirements of listed hazardous chemicals include:
   a. Amounts released or disposed on-site or off-site, the quantities from the previous year, and the quantities anticipated for the next two years.
   b. Amounts recycled on-site and sent off-site for recycling, the quantities from the previous year, and the quantities anticipated for the next two years.
   c. Amounts treated on-site and sent off-site for treatment, the quantities from the previous year, and the quantities anticipated for the next two years.
   d. Amounts used for energy recovery on-site and sent off-site, quantities from the previous year, and the quantities anticipated for the next two years.
   e. Types of source reduction practices implemented and the techniques used to identify those practices.
   f. Methods of recycling used on-site.
   g. Production ratio or activity index to track changes in the level of economic activity at a facility.
   h. Amount of releases resulting from one-time events not associated with production processes.
3. Toxic Release Inventory data for 1992 indicate that 23,630 facilities released 3.2 billion lbs of toxic chemicals into the environment, including air, water, and waste to land.
4. Requirements:
   a. By December 31, 1993, federal agencies were to have provided the Environmental Protection Agency (EPA) with a list of their facilities.
   b. The first federal agency baseline year for reporting chemical releases to the environment of Section 313 Toxic Chemicals was 1994 and included those chemicals listed and in effect on January 1, 1994, including the 34 chemicals added to the list in 1993. One report (on EPA Form R) is required from each facility for each chemical meeting the use criterion. If a facility meets the use threshold, but releases nothing, the facility must still submit a report.

c. Federal facilities having an extremely hazardous substance at or above the threshold planning quantity were to have notified the State Emergency Response Commission and Local Emergency Planning Committee of its presence by March 3, 1994.

d. If a federal facility is required to maintain a Material Safety Data Sheet (MSDS) for a hazardous chemical, and the facility has that chemical at or above the threshold for reporting, the facility is required to submit the MSDS to the State Emergency Response Commission, the Local Emergency Planning Committee, and the fire department with jurisdiction over the facility.

e. If a federal facility has a release of an extremely hazardous substance at or above the reportable quantity, the facility is required to immediately notify the State Emergency Response Commission and the Local Emergency Planning Committee for the areas likely to be affected by the release.

f. The PPA, in Section 6607, requires an annual toxic chemical source reduction and recycling report from each facility required to file an annual toxic chemical release (EPA Form R). The report includes the quantity of the chemical entering any waste stream or the environment, the amount of the chemical recycled, source reduction practices used, the projected amount expected to be reported for the next two years, the techniques used to identify source reduction opportunities, the amount of the toxic chemical released as a result of a catastrophic event, and the amount of the chemical treated at the facility or elsewhere.

g. All federal agencies must complete an agency pollution prevention strategy in writing by August 3, 1994.

h. Each federal agency shall establish a plan and goals for eliminating or reducing the unnecessary acquisition of products containing extremely hazardous substances or toxic chemicals by August 3, 1994.

i. Federal agencies should complete the review of standardized documents to identify opportunities to eliminate or reduce requirements for extremely hazardous substances or toxic chemicals by August 3, 1995.

j. Beginning October 1, 1995, each federal agency must submit an annual progress report to the EPA to include the status of the pollution prevention strategy; efforts to involve the public; the status of facility pollution prevention plans; progress toward the 50 percent reduction goal; progress toward the acquisition and procurement goals; progress in reviewing and revising standardized documents; a sampling of new and innovative pollution prevention technologies fostered; and the total of toxic chemical releases reported for the previous year.

k. Written pollution prevention plans must have been completed for each of a federal agency's "covered facilities" by December 31, 1995. Plans do not have to be submitted to EPA or state agencies, but they should be available if requested by the EPA. Once plans are completed, they are available to the public.

l. For agencies with facilities meeting toxic release inventory chemical reporting thresholds, the agency must develop voluntary goals to reduce the agency's total releases to the environment and off-site transfers of toxic chemicals for treatment and disposal by 50 percent by December 31, 1999. The term "voluntary" goal emphasizes that the 50 percent reduction is a Presidential "goal" rather than an absolute "requirement." The 50 percent reduction goal is not to be applied on a facility-specific basis, but is measured as the "aggregate amount" reported for all of the agency's covered facilities.

m. In considering whether a reporting threshold has been exceeded, toxic chemicals used for any of the following purposes are not included [40 CFR 372.38 (c)]:
   • As a structural component of the facility.
   • In routine janitorial or facility grounds maintenance.
   • In facility operated cafeteria.
   • In motor vehicle maintenance.
   • In process water and non-contact cooling water.

n. There is no requirement to measure or monitor releases for purposes of Section 313 reporting. Readily available data may be used or if data are unavailable, the law allows the reporting of reasonable estimates.

o. If the use of a toxic chemical during a calendar year is less than 10,000 lbs, no reporting is necessary. However, once a report is filed, successive annual reports must be filed even though the reported release is zero.

p. EPA Form R requires the following information for each listed toxic chemical, which exceeds the reporting threshold amount, otherwise used at a facility in yearly amounts:
   • The name and location of the facility.
   • The identity of the listed toxic chemical.
   • The maximum quantity of the toxic chemical on-site at any time during the year.
   • The total quantity of the toxic chemical released during the year, including both accidental spills and routine discharges or emissions.
   • Off-site locations to which toxic chemical wastes were shipped and the quantities shipped.
   • The total quantity of the toxic chemical entering waste prior to recycling, treatment, or disposal.
   • Source reduction activities and other pollution prevention data involving the toxic chemical.

8.4   ENDANGERED SPECIES ACT

1. The applicable sections of the Endangered Species Act include:
   a. Section 4 provides for the listing of a threatened or endangered species.
      • Endangered species means any species which is in danger of extinction throughout all or a significant portion of its range other than a

species of the Class Insect as determined by the Secretary to constitute a pest whose protection under the provisions of this Act would present an overwhelming and overriding risk to man.

• Threatened species means any species which is likely to become an endangered species within the foreseeable future throughout all or a significant portion of its range.

b. Section 4 further provides for a cooperative agreement with a state wherein the state agency is authorized to establish an endangered species program and provision is made for the state to designate resident species of fish or wildlife as endangered or threatened under the state program.

2. Section 7 provides for consultation with the Secretary of the Interior:

a. Each federal agency shall, in consultation with and with the assistance of the Secretary, ensure that any action authorized, funded, or carried out by such agency is not likely to jeopardize the continued existence of any endangered or threatened species or result in the destruction or adverse modification of habitat of such species.

b. A federal agency shall consult with the Secretary on any prospective agency action if the permit or license applicant has reason to believe that an endangered or threatened species may be present in the area affected by a project and that implementation of such action will likely affect such species. If the Secretary advises that such species may be present, such agency shall conduct a biological assessment for the purpose of identifying any endangered or threatened species which is likely to be affected by the project.

3. Section 9 lists a number of prohibited acts:

a. To take a listed endangered or threatened species is prohibited.

b. The term "take" means to harass, harm, pursue, hunt, shoot, wound, kill, trap, capture, or collect, or to attempt to engage in any such conduct.

4. Section 10 provides for the issuance of a permit for the incidental taking of an endangered species. However, no permit may be issued unless the applicant specifies.

a. The impact that likely will result.

b The steps to be taken to minimize and mitigate impacts.

c. Funding available to implement mitigation actions.

d. Potential alternative actions to the takings.

e. Other measures as may be appropriate.

8.5   MARINE MAMMAL PROTECTION ACT OF 1972

1. The applicable sections of the MMPA include:

a. Section 102 (a) states that it is unlawful for any person subject to the jurisdiction of the United States or any vessel or other conveyance subject to the jurisdiction of the United States to take any marine mammal on the high seas.

b. Section 101 (a) (5) (A) allows the incidental, but not intentional, taking by citizens with a five-year permit issued after notice in the *Federal Register* with an opportunity for public comment, and a finding that the total of such taking during each five-year period will have a negligible impact on such species or stock and will not have an unmitigable adverse impact on the availability of such species.

c. Section 101 (a) (5) (D) allows a one-year authorization for the incidental, but not intentional, taking by harassment of small numbers of marine mammals. The authorization will only be allowed after a finding that such harassment will have negligible impact and will not have an unmitigable adverse impact on the availability of such species.

d. Section 3 defines a "person" to include any officer or employee of the federal government. The term "take" means to harass, hunt, capture, or kill, or attempt to harass, hunt, capture, or kill. The term "harassment" means any act of pursuit, torment, or annoyance which has the potential to injure or disturb a marine mammal by causing disruption of behavioral patterns, including feeding.

e. Section 104 (h) provides that general permits may be issued for the taking of marine mammals, together with regulations to cover the use of such general permits.

2. The legislative history provides information on the purpose and intent of the legislation.

a. House of Representatives (HR) Report No. 92-707 states that the purpose of this legislation is to prohibit the harassing, catching, and killing of marine mammals. Recent history indicates that "whales, porpoises, seals, sea otters, polar bears, and manatees have been shot, blown up, clubbed to death, run down by boats, poisoned, and exposed to a multitude of indignities" (page 12). "Manatees and sea otters have been crippled and killed by motorboats and at present the federal government [has been] essentially powerless to force these boats to slow down" (page 15). As many as "200 to 400 thousand porpoises per year were killed by tuna nets" (page 15). The definition of taking includes the concept of harassment and it is intended that this term be construed sufficiently broadly to allow the regulation of "wanton" operation of powerboats (page 18).

b. Senate Report No. 92-863 states that the principles of resource protection and conservation embodied in this Act must be maintained (page 7).

c. HR Report No. 95-336 states that a large part of the time, money, and effort of the Marine Mammal Commission and the Department of Commerce has been spent in an attempt to find a solution to the tuna-porpoise problem (page 4).

d. HR Report No. 97-228 restated the purpose of the MMPA as ensuring that marine mammals are maintained at healthy population levels. It is stated that the Act has been remarkably successful in that the incidental take of porpoises in tuna fishing operations has dropped from 368,000 in 1972 to an estimated 15,303 killed in 1980 (page 11).

e. Senate Report No. 103-220 states th;at whale-watching operations have not been required to apply for permits because they traditionally have had little interaction with the marine mammals under observation (page 5).

f. Senate Report No. 103-220 reviews the federal management of marine mammals. The Department of Commerce through the National Oceanic and Atmospheric Administration has authority with respect to whales, porpoises, seals, and sea lions. The Department of the Interior through the Fish and Wildlife Service has authority with respect to walruses, polar bears, sea otters, and manatees. In carrying out these responsibilities, both agencies are required to consult with the Marine Mammal Commission, an independent advisory agency created by the MMPA. The Commission consists of three part-time Commissioners who must be experts in marine ecology and is supported by a permanent staff of about 10 people (page 2).

3. A citizen's suit (Richard Max Strahan of GreenWorld, Inc., vs. the U.S. Coast Guard) was filed on complaint that two Northern Right whales, Eubalaena glacialis, were killed owing to collisions with Coast Guard vessels. One death occurred in 1991 and the other occurred in 1993. On July 6, 1991, the Coast Guard Cutter "Chase," traveling at 20 to 22 knots, struck and killed a Northern Right whale off the Virginia Cape, southeast of the Delaware coast. On January 5, 1993, the Cutter "Point Francis," which allegedly was traveling at full speed, struck and mortally injured a juvenile Northern Right whale, which was found on shore a few days later and examined by the U.S. Fish and Wildlife Service. The two strikes did not occur during search and rescue or law enforcement operations. The location of the 1993 strike was not identified.

a. The suit was filed under the National Environmental Policy Act, Endangered Species Act, Marine Mammal Protection Act, and Whaling Convention Act.

b. The Court found that this case presents a "challenge to neglect" by the U.S. Coast Guard in that the Coast Guard has not yet applied for an Endangered Species Act incidental take permit, a Marine Mammal Protection Act small take permit, nor completed an environmental assessment as directed by the National Environmental Policy Act.

• The Coast Guard stated that the National Marine Fisheries Service likely could not issue a small take permit for the Northern Right whales because the depleted nature of the stock would prevent a finding that such takes would have a negligible impact upon the species.

• Further, the Coast Guard stated that since incidental takes may be avoided through the Endangered Species Act Section 7 consultation process, application for a small take authorization would be unnecessary.

c. The Court found that the Coast Guard has taken certain actions since the Complaint was filed on June 7, 1994. These actions include:

• Forming an Endangered Species Act Compliance Team.

- Forming an Endangered Species Act Biological Assessment Team.
- Contributing $80,000 toward aerial monitoring of the Northern Right whale calving season from December 1, 1994 to March 31, 1995.
- Requiring Coast Guard vessels to have a designated lookout to assist the conning officer or coxswain in watching for other marine traffic and in avoiding whales and turtles in the water.

d. The Court found that "The Endangered Species Act is a powerful and substantially unequivocal statute." Section 9 (a)(1)(B) makes it unlawful for any person to take any endangered species of fish or wildlife within the territorial sea. However, the Secretary of Commerce is authorized, after an opportunity for public comment, to issue permits authorizing any taking otherwise prohibited by the Act. A detailed conservation plan must be submitted by the applicant for a permit.

e. The Court ordered the following:
- The defendant shall not seek or acquiesce in any extension to the time frames established by the Endangered Species Act Section 7 (b) without prior approval by the Court.
- On October 2, 1995, the defendant shall submit a status report to the Court detailing its intent to proceed with actions in light of the agency's obligations under the Endangered Species Act and the Marine Mammal Protection Act.
- "The Coast Guard is required to apply for a small take permit if it anticipates that it will take a marine mammal at any time during the course of its operations." The Coast Guard is directed to apply for a small take permit from the National Marine Fisheries Service for all Coast Guard operations that may accidentally take a Northern Right whale, regardless of whether defendant considers the possible taking unlikely.
- The Coast Guard is directed to record incidents involving whale strikes by Coast Guard vessels and promptly report any resulting injuries of Northern Right whales to the Court and to the National Marine Fisheries Service.
- By June 30, 1995, the Coast Guard shall submit to the Court a copy of the draft Environmental Assessment of Coast Guard ship operations and their potential environmental impacts, and an anticipated schedule for completion of a final Environmental Assessment and an Environmental Impact Statement or a "Finding of No Significant Impact," whichever is applicable. The schedule shall incorporate the public involvement directives of NEPA.

8.6   FEDERAL INSECTICIDE, FUNGICIDE, AND RODENTICIDE ACT (FIFRA)

1. Regulates the manufacture, distribution, sale, and use of pesticides to minimize risks to human health and the environment.

    a. Defines pesticide as any substance intended to prevent, destroy, repel, or mitigate any pest.

    b. Requires registration of all pesticides, restricts use of certain pesticides, establishes requirements for certification of applicators, authorizes experimental use permits, creates conditions for registration cancellation, requires registration of manufacturers, and sets standards for pesticide disposal.

    c. Regulatory decisions are made by balancing of risks and benefits, i.e., a pesticide must perform its intended function without causing unreasonable adverse effects on human health or the environment when balanced against the potential benefits of the proposed use.

    d. Manufacturers of pesticides are required to provide data on the potential for skin and eye irritation; hazards to non-target organisms, including fish and wildlife; the possibility of acute poisoning, tumor formation, birth defects, reproductive impairments, or other serious health effects; the behavior of the chemical in the environment after application; and the quantity and nature of residues likely to occur in food or feed crops.

    e. Under the provisions of the Federal Food, Drug, and Cosmetic Act, the EPA established tolerances for pesticide residues on feed crops and raw and processed foods. Tolerances are established at levels below amounts that might cause harm to people or the environment. For agricultural commodities, tolerances are enforced by the Food and Drug Administration. In meat, poultry, and fish products, tolerances are enforced by the U.S. Department of Agriculture.

2. Section 3 prohibits the distribution and sale of pesticides that have not been registered with the EPA and establishes procedures and data requirements for registration. The section also provides the requirement for classification of pesticides for general use, restricted use, or both.

    a. General use pesticides are considered safe for use by anyone, provided label directions, restrictions, and precautions are observed.

    b. Restricted-use pesticides may be used only by persons who have been certified as trained applicators. Training and certification is administered through EPA-approved state pesticide programs.

    c. The EPA considers three types of pesticide applications:
- A pesticide containing an active ingredient that is not a constituent of a product currently registered.
- An application for registration of a use for an active ingredient not currently included in the directions for use of any product that contains such active ingredient.
- One that is substantially similar or identical in its uses and formulation to products that are currently registered.

    d. For a new active ingredient that has not been marketed before, it may take 6 to 9 years and $2 to 4 million to comply with all registration data requirements. The EPA may issue experimental use permits or temporarily authorize state or federal agencies to combat emergencies with pesticide uses not permitted by existing federal registrations.

    e. Applicants who use data submitted by another in support of their own application for registration must compensate the original data submitter for use of the data.

3. Section 4 requires reregistration of each registered pesticide containing any active ingredient first registered before November 1, 1984 to ensure that data requirements of current law are in place and that "when used in accordance with widespread and commonly recognized practice it will not generally cause unreasonable adverse effects on the environment."

4. Section 5 allows issuance of an experimental use permit to enable a manufacturer to develop data necessary to register a pesticide under Section 3.

5. Section 6 provides for the registration of any pesticide to be canceled at the end of five years unless the registrant requests that the registration be continued in effect.

6. Section 13 authorized the EPA to issue an administrative order to stop the sale, use, or removal of any pesticide that is reasonably believed to be in violation of the Act; has been, or is intended to be, distributed or sold in violation of the Act; or has been canceled or suspended. The section also authorizes seizure of any pesticide that has been adulterated or misbranded, has not been registered; bears inadequate or improper labeling; has not been colored or discolored if required; differs in its claims or use directions compared to those in the registration application; or causes unreasonable adverse environmental effects when used in accordance with applicable requirements and restrictions.

7. If a registered pesticide shows evidence of posing a potential safety problem, the EPA can conduct a "special review" of risks and benefits in which all interested parties can participate. Depending upon findings, the agency may implement regulatory options, including cancellation or suspension proceedings, restricting pesticide use to certified applicators, requiring protective clothing when applying the chemical, prohibiting certain application methods, prohibiting certain uses, or continuing the registration.

8.7 FISH AND WILDLIFE COORDINATION ACT

1. Provides that wildlife conservation shall receive equal consideration and be coordinated with other features of water-resource development programs.

2. Whenever the waters of any stream or other body of water are proposed or authorized to be impounded, diverted, the channel deepened, or the stream or other body of water otherwise controlled or modified for any purpose whatever, including navigation and drainage, by any department or agency of the United States, or by any public or private agency under federal permit or license, such department or agency first shall consult with the U.S. Fish and Wildlife Service and with the applicable state agency.

8.8   MIGRATORY BIRD TREATY ACT

This act protects migratory birds, their nests, and eggs from being hunted, captured, purchased, or traded. If pesticides are used to manage bird populations other than starlings, sparrows, and pigeons, a permit is required for their taking from the U.S. Fish and Wildlife Service.

8.9   BALD EAGLE PROTECTION ACT

The Bald Eagle Protection Act provides for the protection of bald and golden eagles.

8.10   NATIONAL HISTORIC PRESERVATION ACT

The act requires an expanded National Register of Historic Places and establishes the Advisory Council of Historic Preservation. Section 106 of the Act requires federal agencies to allow the Advisory Council an opportunity to comment whenever their undertakings may affect National Register resources or resources that are eligible for listing on the National Register. Section 110 of the Act requires federal agencies to identify, evaluate, inventory, and protect National Register resources on properties that they control. The Act imposes no absolute preservation requirements, as long as mandated procedures are followed and documented in any federal action that may impact National Register resources or resources that are eligible for listing on the National Register.

8.11   COASTAL ZONE MANAGEMENT ACT OF 1972

1. Provides that the Secretary of Commerce, following management program approval, may make matching fund grants to any coastal state for the purpose of administering that state's coastal zone management program.
2. Before approving a management program, the Secretary shall find that the management program includes:
   a. An identification of the boundaries of the coastal zone subject to the management program.
   b. A definition of what shall constitute permissible land uses and water uses within the coastal zone.
   c. An inventory and designation of areas of particular concern within the coastal zone.
   d. An identification of means by which the state proposes to exert control over the land uses and water uses.
   e. A description of the organizational structure.
3. The coastal zone management program generally incorporates flood control, sediment control, grading control, and stormwater runoff.
4. Coastal zone is defined as the coastal waters and adjacent shorelands.
   a. The zone extends inland from the shorelines only to the extent necessary to control shorelands, the uses of which have a direct and significant impact on the coastal waters.

b. Coastal waters in the Great Lakes means the waters within the territorial jurisdiction of the United States consisting of the Great Lakes, their connecting waters, harbors, roadsteads, and estuary-type areas such as bays, shallows, and marshes. Coastal waters in other areas means those waters, adjacent to the shorelines, which contain a measurable quantity or percentage of sea water, including, but not limited to, sounds, bays, lagoons, bayous, ponds, and estuaries.

9 Compliance

Compliance is the degree of meeting the environmental regulations and laws applicable to a facility's activities. Determining compliance requires a knowledge of the facility's activities, and the laws and regulations pertinent to it. Such knowledge must include the requirements of all permits issued to the facility, as well as the monitoring and reporting requirements contained in the permits. Generally, audits are used to determine the degree of compliance. To avoid failing an audit by a regulatory entity, self-audits on a routine or annual basis are helpful to uncover and mitigate any potential compliance violations before a regulatory audit occurs.

When there is knowledge of a violation of an environmental law or regulation as a result of a self-audit, steps should be taken to correct any deficiency. A file of those correction steps should be maintained in written documentation. Self-audit reports or results can become available to regulatory authorities.

The objective of an audit is to establish the degree of compliance with applicable laws, regulations, permits, and good management practices related to an operation. An audit provides management with knowledge of the degree of compliance, a list of needs or adjustments necessary to attain full compliance, and insight into the staff's ability to provide information in a timely fashion. Generally, the conduct of an audit entails a solicitation of responses to questions designed to uncover inadequacies in the system, an evaluation of responses and other information obtained as a result of the questioning, and a statement of findings of fact and needs for compliance.

An audit of hazardous waste operations, e.g., might include some of the following questions. It is good practice to provide the reference or requirement for an audit question to facilitate finding the regulatory requirement. See Table 9.1.

TABLE 9.1
Sample Questions for a Hazardous Waste Management Self-Audit

- Does the facility generate, store, transport, treat, or dispose of a hazardous waste?
- What are the assigned federal and state identification numbers? (40 CFR 262.12)
- Describe and identify by hazardous waste number the hazardous wastes generated. (40 CFR 261.3)
- List and describe the manufacturing processes producing the waste.
- Provide schematic diagrams of all such manufacturing processes.
- Provide a complete list of raw materials used in the manufacturing process.
- Identify and describe any other substances used in the manufacturing process, including solvents, cleaners, degreasers, and coating or painting materials; indicate maximum weekly usage. [40 CFR 261.3 (a)(2)(iv)(A)]
- What is the average weekly flow of wastewater into the headworks of the facility's wastewater treatment or pretreatment system?

- Provide material safety data sheets for all substances used.
- What are the average and maximum volumes of waste generated per month and per year? (40 CFR 261.5)
- Describe any waste treatment system connected with hazardous waste management.
- Provide a schematic diagram of the waste treatment system.
- Provide a general description of the hazardous waste management facility.
- What were the disposal methods and where was disposal undertaken for wastes generated prior to November 1980?
- Provide a copy of the laboratory testing results related to the facility's hazardous waste.
- Have there been any process or product changes that would alter the waste since the last laboratory results?
- Provide a copy of the laboratory's quality assurance program and procedures.
- What has been the procedure for disposing of unused commercial chemical products? (40 CFR 261.33)
- What has been the practice for disposing of empty containers that previously held hazardous waste? (40 CFR 261.7)
- Describe any packaging, labeling, or marking of hazardous waste prior to its transportation.
- Provide a file of the generator's copies of transportation hazardous waste manifests. (40 CFR 262.21)
- Provide a file of the generator's annual reports. (40 CFR 262.41)
- Provide a file of the generator's exception report file. (40 CFR 262.42)
- How long are hazardous wastes accumulated on site? (40 CFR 262.34)
- Do you transport your own hazardous waste to the disposal site? What are the necessary actions relating to spills during transportation? (40 CFR 263.30 and 263.31)
- Provide a copy of the facility's waste analyses plan. [40 CFR 265.13 (b)]
- What security measures are provided to prevent unauthorized entry or human or other animal contact with the waste? (40 CFR 265.14)
- What security measures are provided to protect employees from areas where hazardous wastes are stored? (40 CFR 265.14)
- Provide a copy of the facility's inspection schedule and file on any inspections performed. (40 CFR 265.15)
- Describe the program for personnel training. (40 CFR 265.16)
- Have the police, fire department, and emergency response teams been made familiar with the layout and the properties of the hazardous waste? (40 CFR 265.37)
- Provide a contingency plan to minimize hazards to humans. (40 CFR 265.51)
- Have copies of the contingency plan been submitted to local police, fire, and other emergency officials? (40 CFR 265.53)
- Who is the identified emergency coordinator for the contingency plan? (40 CFR 265.55)
- Provide a file of operating records for the facility. (40 CFR 265.63)
- Describe container or tank storage management practices.
- Provide plans for container or tank storage areas.
- Provide a Best Management Plan for controlling runoff and leachates from the plant site.
- Provide a SPCC plan for the container-tank storage and other appropriate areas.
- Has the groundwater sampling and analyses program been implemented? Provide a copy of the laboratory results.
- Provide an outline of the groundwater quality assessment program. [40 CFR 265.92 (a)]
- Provide a plan for the facility's closure. (40 CFR 265.112)
- Have adequate financial assurances for closure and postclosure operations and liability protection been provided?

- Provide the employee accident report file.
- Provide the citizen complaint file.
- Provide the federal-state comment-correspondence files.
- Describe the employee supervision structure.
- Are the employees provided job descriptions? Does each employee have a copy of his/her job description?
- Provide a copy of the hazardous waste transporters' Certificates of Liability Insurance.
- Provide a copy of logs of inspections of above, surface, or underground hazardous waste containers or storage facilities.

Recently the EPA, in cooperation with nine other federal agencies formed an Interagency Environmental Audit Protocol Workgroup for the federal community and produced a comprehensive two-volume book entitled *Generic Protocol for Conducting Environmental Audits of Federal Facilities*. The contents of the volumes can be adapted for industrial use as well. The books contain appropriate and comprehensive audit questions aimed toward bringing federal facilities into compliance with environmental laws and regulations related to facility operations. The volumes provide a useful guide to auditing activities.

The previously described protocol addresses the following 16 program areas. Audit questions are provided on each of the 16 areas. The document is a helpful reference.

Air pollution control
Water pollution control
Non-hazardous waste management
Hazardous waste management
CERCLA/SARA
Spill control and response
Management of environmental impacts (i.e., NEPA and NEPA components)
Hazardous materials management
Emergency planning and community right-to-know
Cultural and historic resources management
Storage tank management
Drinking water management
PCB management
Pesticide management
Groundwater protection
Environmental radiation

10 Ecosystem Recovery

How long after pollution has been controlled will the water be fit for swimming, fishing, or aesthetic enjoyment? Ecosystems do heal and the healing is a very noticeable thing. How rapidly this occurs depends on a number of interacting factors that contribute to recovery. Recovery, also, is different for each of the three media. The air media can recover rapidly because air currents move and disperse pollutants from a given area. The land media recovers very slowly because recovery may depend upon the growth of vegetation. In the water environment, each waterway is a unique microcosm comprised of physical, chemical, and biological interacting factors.

Theoretically, the recovery of a waterway that is not influenced significantly by tributaries or other similarly acting factors, and from which a significant source of pollution has been removed, should proceed toward recovery with a shortening of the pollution zones. The downstream or horizontally located zone of recovery should move upstream and gradually replace the zone of active decomposition. Eventually a similar process should occur in the zone of degradation. The adjacent downstream or more distant clean-water zone likewise should follow and eventually replace the zone of recovery when the effects of the insult on the aquatic environment have been obliterated. A succession of organisms would occur in the polluted area beginning with the bacteria and associated protozoa and algae and extending through the macrobenthos, which would include the sludgeworms, midge larvae, fingernail clams, sowbugs, leeches, dragonfly nymphs, caddisfly larvae, and other gill-breathing insect larvae in that order. Eventually a population of fish and other aquatic life would be indistinguishable in population characteristics from an area of the waterway that remained unaffected by pollution. Often the events in nature do not occur in this logical sequence, however, because of the interaction of the many factors that will be discussed later. It is this process of interaction that complicates the predictability of natural phenomena that in combination produce waterway recovery.

Significant influencing factors in the waterway recovery process include the extensiveness of environmental damage, the length of time that the environment has been degraded, the environments affected, and the possibilities for infiltration of biota.

The extensiveness of environmental damage to a particular waterway habitat is different for each cause of environmental destruction. Such damage may be subtle and may affect only one or a few species of the most sensitive organisms within the aquatic community. The other extreme would be virtual destruction of the community of plants and animals within a given reach of waterway. An innumerable number of possibilities for different environmental gradations exist between these two extremes.

The cause of the environmental destruction may have been continuous, intermittent, fluctuating in degree, seasonal, annual, or occurring once in 10, 50, or 100 years. Each action will produce a different reaction within the aquatic community.

The history of environmental degradation or the length of time that such degradation has occurred will tend to affect the length of time required for recovery. For example, a lake that has become eutrophic within the past decade, and within which the process has been a gradual one, will recover in a much shorter time span that a lake with a long history of eutrophication.

The types of waterway environments affected by a particular catastrophe will affect the time required for recovery when the catastrophe subsides or has been controlled. The flowing water environment will react differently from that of the non-flowing water environment.

When pollution has been removed or controlled in a flowing stream, recovery often is aided by certain natural phenomena that occur seasonally or at more or less defined intervals. A spring flood, for example, will remove sludge and silt deposits or dilute and wash away the residual pockets of toxic materials that may remain from a source of toxic pollution. This cleansing effect will prepare a physical habitat that is more suitable for colonization and inhabitation by organisms than one associated with a polluted environment. If nutrients have been a problem in the stream environment, these too are diluted and cleansed from the area by the forces and actions of spring freshets. Should the expected spring floods fail to occur because of prevailing area drought conditions, the time of recovery for a flowing water environment may well be extended for one or two additional years.

Expected recovery in the non-flowing environment generally will require a longer period of time than expected recovery within the flowing environment. Sludge deposits and their resultant environmentally deleterious effects are not easily removed by floods or spring freshets. Quite often the effects of such deposits must be minimized through the natural blanketing of such areas with naturally occurring materials that are environmentally more acceptable. The covering-over process is a much slower one than the attrition caused by flooding and may require several years to complete. In the interim, biological degradation of the sludge banks continues with some deleterious environmental effects that are diminished progressively with the healing influence of time.

Recovery within the non-flowing environment depends to a large extent on the flow-through time or water retention time of the lake or reservoir. For example, the flow-through time for Lake Michigan is approximately 15 times the flow-through time for Lake Erie. The theoretical recovery time following a similar degree of abatement of a similar pollutional load would approximate a ratio of 15 to 1 with Lake Erie experiencing the more rapid recovery. The physical topography of a water basin affects its retention time and significant factors include the length, width, depth, contour of the bottom, physical location of inflowing and outflowing streams, volume of water within the holding basin compared to the annual volume of inflowing water, water temperature, and other related factors.

Probably the most significant pollution problem associated with lakes is that of eutrophication. All natural lakes through geologic time tend to mature, become more fertile, eventually fill with organic materials, and ultimately return to the landscape as dry land. This aging process has been accelerated in a great number of lakes through the activities of man. The nutrients, principally in the form of available nitrogen and phosphorus, that reach a lake basin through discharges of pollution to the basin or to the drainage area result in increased algae and vascular plant nuisances. These have contributed to an accelerated accumulation of organic materials. The general process associated with a lake becoming overenriched before its geologic time has been referred to as cultural eutrophication.

Once nutrients are removed from the lake or its drainage basin through processes of advanced waste treatment or other controls, lake recovery will begin. It may not ever be possible to bring the condition of a lake back to a point on the geologic time scale when the cultural eutrophication process was initiated. It should be possible, on the other hand, to recover a waterway to such an extent that many valuable decades of multiwater use can be obtained for a citizenry that has displayed an innate affinity for association with the water environment.

Another significant factor influencing recovery is the possibilities for infiltration of organisms into the organism-depleted environment. These possibilities include downstream drift, upstream migration, general invasion from adjacent areas as in a lake, egg deposition of those aquatic organisms that have terrestrial adults, infiltration by wind-borne algae and other organisms, transportation of aquatic plants, algae, and microorganisms by birds and mammals, and the influences of man in transporting and transplanting aquatic organisms intentionally or unintentionally, into an otherwise denuded area.

When the recruitment of organisms from adjacent or upstream areas does not occur, the principal process of establishing a biotic community within an area where the previous community has been virtually destroyed is through ecological succession. In the process of succession, bacteria from the soils and those introduced from the atmosphere convert the organic materials present to protoplasm and serve as a food supply for protozoa and other microscopic organisms. Simultaneously, minute algae contribute similarly to the development of an environment that will be amenable to the life processes of higher and more complex biotic forms. When the environment attains a quality that will support their existence, other organisms will appear. Each type of organism in turn, will influence the aquatic environment and prepare it for the invasion of a higher form of life, which culminates in a balanced fish population. In an undisturbed aquatic environment the predator fish will assume the role of dominant organisms and will control the biotic population to the advantage of their species.

The type of combination or recruitment activities to fill a biotic environmental void is an important determinant of the time interval required for recovery. The downstream drift of organisms from upstream riffle areas or from tributaries may attain rates of many pounds per day and may serve to repopulate an area rapidly providing conditions for existence are acceptable for the survival of the species. The downstream drift of organisms is a common daily occurrence, but there is evidence that the greatest drift may occur during the non-daylight hours. Downstream drift is common among insect larvae and occurs when the immature forms become released from their points of attachment and drift with the current.

Upstream migration of organisms is another phenomenon that is not as effective as downstream drift in the repopulation of an area, but, nevertheless, is a significant factor. Evidence indicates that the amount of downstream drift far exceeds the amount of upstream migration for a given stream reach. In a lake environment, repopulation can be aided to a great extent from a general invasion by organisms from adjacent lake areas. Presumably, the invasion would be from all accessible sides and would occur when conditions for existence would support the organisms in accordance with their general adaptability for locomotion.

Egg deposition can be a significant means of repopulating an area particularly by those species of organisms that have a terrestrial stage. When the chemical water quality is returned to an acceptable level for the survival of immature insect forms, the forms usually will be present within the environment following the time of the next egg-laying period for the species.

Thus, a flowing water area would tend to become repopulated with invertebrates within a maximal period of one year following a corrected or controlled catastrophic occurrence. The interval of time might be decreased depending upon the possibilities of downstream drift, upstream migration, or general invasion. The lake or pond environment is more complex and depends upon a number of related factors, and the recovery period probably would be longer.

Algae and other microscopic organisms are found in the atmosphere and may be transported over great distances by the winds. This is a major reason why algae may be found in practically every habitat known to man. Birds are a significant transporter of fragments of aquatic vascular plants, algae, and other microscopic organisms. Such organisms may be carried in the intestinal tract of birds or as passengers on feet or beaks. When birds alight upon the water, such organisms are dislodged and can mature in the new environment. Certain mammals also, particularly those that are associated with the aquatic environment, can transport algae, aquatic vascular plants, and other aquatic microorganisms. Man himself assumes an important part by the practice of intentionally or accidentally planting or introducing various species into aquatic habitats. Often such species, unencumbered by natural enemies, become a severe pest and cause the expenditures of large sums of money aimed toward their eradication or control. Notable examples of such introductions that have become significant pest species include the german carp, Eurasian water milfoil, zebra mussel, and the Asiatic clam. The occurrence of certain other species in southern waters of the United States that are pests in more tropical climes is a matter of grave concern and current investigation.

The recovery of a lake from the degrading effects of eutrophication is a slow process of attrition of the nutrient supply. Bottom sediments may become covered with inert silts that serve to segregate the organic materials from the superimposed water. In time, the nutrients that are recycled within the biomass decrease. With the death and decomposition of organisms, a portion of the nutrients combine with the consolidated benthic sediments and do not recharge to the superimposed water. Nutrients are discharged continually in the lake's effluent and here the flow-through time of the water in the lake basin is an important factor in determining the time necessary for lake recovery. When the flow-through time is a few months or two or three years, the recovery rate from eutrophication may be a decade or less. When the flow-through time is counted in decades or many years, the recovery period will be longer.

11 Pollution Prevention

The two words that currently have captured center stage in environmental considerations, at least in federal agencies, are Pollution Prevention. As defined in the Pollution Prevention Act, pollution prevention means source reduction and other practices that reduce or eliminate the creation of pollutants through:

1. Increased efficiency in the use of materials, energy, water, or other resources.
2. Protection of natural resources by conservation.

Prevent pollution at the source to eliminate or minimize adverse health effects while protecting, preserving, restoring, and enhancing the quality of the environment. The Pollution Prevention Act of 1990 identifies an environmental management hierarchy in which pollution "should be prevented or reduced whenever feasible; pollution that cannot be prevented should be recycled in an environmentally safe manner whenever feasible; pollution that cannot be prevented or recycled should be treated in an environmentally safe manner whenever feasible; and disposal or other release into the environment should be employed only as a last resort" (42 U.S.C. 13103).

The development and implementation of a pollution prevention plan is required for all facilities that meet or exceed the threshold for reporting and filing an EPA Form R for releases to the environment pursuant to EPCRA. The development and implementation of a facility pollution prevention plan can take on a much broader concept, however. The plan can encompass nearly all activities associated with operating a facility. It can prevent pollution, conserve resources, and save operating money. From the cost savings standpoint alone, it is well worth looking into.

To be effective, a pollution prevention program should be in the minds of all employees. Thus, the plan for the program should originate with the employees through an employee committee where everyone participates in suggesting activities where savings in resources and labor can occur. These ideas should be recorded and become part of the plan as it is developed. Search for ways to do a job easier, more efficiently, with less labor, and a savings in resources. Get in the habit of maintaining a file on activities that could fit into the preceding concept. If the time is not ripe to implement such measures now, the file may be valuable at a later time.

Minimization incentives should be implemented to heighten the awareness of all staff through training and acquisition activities. Such incentives can result in reduced potential pollutant uses through a program that:

- Incorporates waste minimization modules into all relevant training regimens.

- Implements waste minimization objectives in the systems acquisition life cycle.
- Develops standards for reductions in packaging supplies, and implements them through training and performance requirements.
- Conducts research to identify, develop, and document standard practices that result in pollution prevention and waste minimization.
- Encourages acquisition of long-term use items as opposed to acquisition of one-use throw-away items.

Implementing a successful use minimization program is based on a knowledge of:

- Potential pollutants being used in a life-cycle process.
- Repetitive activities wherein savings in time, labor, or resources could accrue if it were possible to find a better way of completing the task.
- Potential alternatives for use with decreased adverse environmental and resource effects.

Personnel training is vital to the success of any program. Implement a personnel training program. The following could be the guiding principles:

- Protect the environment.
- Conserve natural resources.
- Reduce waste; prevent pollution.
- Comply with applicable environmental laws and regulations.

Establish priorities:

- Prevent waste generation.
- Minimize hazardous materials introductions.
- Recover waste and recycle to the maximum extent.
- Ensure environmental safety and regulatory compliance.
- Release to the environment only when necessary.

The formal pollution prevention plan should be developed to consider the following headings:

Purpose.
Policy.
Applicability and scope.
Definitions.
Sources of waste.
Plan management.
Source reduction.
Use minimization.
Alternative substitution.
Personnel Training
Priorities for Implementation

12 Review Questions and Answers

12.1 QUESTIONS

1. What important part of a *Federal Register* final rule is not codified in the CFR?

2. What national program does Section 303 of the Clean Water Act address?

3. What is the current OSHA PEL for asbestos in fiber per cc?

4. Name two of the four components of a Water Quality Standard.

5. For what does Section 312 of the Clean Water Act require that standards be promulgated?

6. When (approximately) was the first federal water pollution control legislation enacted?

7. What is the purpose of water quality criteria in water quality standards?

8. Circle the most correct answers. An NPDES permit is needed for:
   a. The discharge of dredge and fill material.
   b. Any discharge of toxic material into water of the United States.
   c. Any emission of ozone.
   d. Any pollutant discharged to waters of the United States.

9. Circle the most correct answer. Permits for dredge and fill material are issued by:
   a. Environmental Protection Agency.
   b. Corps of Engineers.
   c. Fish and Wildlife Service.
   d. Coast Guard.

10. Circle one answer. Federal regulations are promulgated by the Congress.
    True   False

11. Circle one answer. Proposed rules are enforced by the EPA.
    True   False

12. Circle one answer. The District of Columbia area is in attainment for ozone.
    True   False

13. Circle one answer. The District of Columbia area is in non-attainment for carbon monoxide.
    True   False

14. What group of substances are regulated by the NESHAP for Ship Building and Ship Repair?

15. What is the required disposal method for liquids with PCBs equal to or greater than 500 ppm?

16. Circle one answer. PCBs can still be introduced into commerce.
    True   False

17. If you circled "true" in question 16, what is the form in which PCBs can still be introduced into commerce?

18. Circle one answer. TSCA banned certain uses of PCBs in 1976. However, the continued use of pre-1976 PCBs was grandfathered into the law.
    True   False

19. Under TSCA, a manufacturer can use another manufacturer's environmental data for a product to meet permit application requirements, providing compensation is provided to the initial collector of the data.
    True   False

20. What is "friable" asbestos?

21. What three control measures can be employed when severely damaged friable asbestos is found?

22. Under what section of TSCA can the EPA take action to prohibit or restrict the introduction into commerce of a chemical or chemical substance?

23. Conducting a toxicity test, i.e., subjecting specific test organisms to a particular waste concentration, is often a requirement of an NPDES permit.
    True   False

24. Executive orders are signed by the President to require industry to stop polluting.
    True   False

25. Circle all that are appropriate. In developing a draft permit, a permit writer may use:
    a. Effluent guidelines.
    b. Toxicity testing results.
    c. Executive orders.
    d. Best engineering judgment.
    e. Water quality standards.

26. When is a solid waste a hazardous waste?

27. Circle one answer. An identification number is not needed to haul hazardous waste interstate?
    True   False

28. Hazardous waste can be stored on-site without a permit for _____ days.

29. A small quantity generator is one who generates between _____ and _____ kg/month of hazardous waste.

30. Circle one answer. Every permit issued for a discharge to water, of any kind whatsoever, must ensure that water quality standards shall not be violated.
    True   False

31. Small generators of hazardous waste can legally store hazardous waste for _____ days.

32. Circle all that are appropriate: In the disposal of dredged spoil:
    a. Clean Water Act Section 404(b) guidelines govern disposal in the ocean (out to 3 nm).
    b. The Corps of Engineers always has the last word on permit issuance.
    c. The Corps of Engineers may issue a permit for the transportation of dredged material for the purpose of dumping it into ocean waters.
    d. It is never necessary to conduct toxicity tests prior to dredged spoil disposal.

33. The CFR is another way of indexing federal laws.
    True   False

34. The National Priority List is a list of (circle all that are appropriate):
    a. Chemicals that must be analyzed for in an NPDES permit application.
    b. Sites with threatened releases of hazardous substances.
    c. Fast food places along Interstate 70.
    d. Pesticides for intensive environmental study.

35. Circle one answer. The National Contingency Plan provides instructions for the clean up of hazardous waste sites only.
    True   False

36. Circle one answer. Ocean dumping is a viable option for the disposal of sewage sludge from the New York area.
    True   False

37. Circle one answer. It is legal for a Navy ship to discharge floating solid waste in the ocean beyond 25 nm from land.
    True   False

38. Circle one answer. The Coastal Zone Management Act regulates land use in a 25-mi wide parcel of land along the coastline.
    True   False

39. Circle one answer. EPCRA provides credits to industry for pollution prevention planning.
    True   False

40. Circle one answer. FIFRA authorized the issuance of experimental use permits to enable a manufacturer to develop data necessary to register a pesticide.
    True   False

41. Check correct answer. The Federal Facilities Compliance Act provided that public vessels are _____ are not _____ generators of hazardous waste.

42. Circle one answer. An ocean dumping general permit provides for human burial at sea.
    True   False

43. Check correct answer. Industrial wastes have been _____ have not been _____ prohibited by an act of Congress from being dumped into the oceans.

44. Circle one answer. It is now unlawful for vessels, other than public vessels owned or operated by the DOD, to discharge plastics into the oceans.
    True   False

45. Circle the appropriate answers. Under the Endangered Species Act:
    a. To "harass" a listed species is unlawful.
    b. To "pursue" means the same as to "take."
    c. To "watch" closely a listed species is to "harass."
    d. An endangered species is one in "short" supply.

46. Circle one answer. Under EO 12856 and EPCRA, any facility that has to fill out an EPA Form R for the release of a chemical must continue to file the form annually regardless of the quantity of the chemical released.
    True   False

47. The threshold for reporting on Form R by a federal agency generally is _____ lb of a listed EPCRA chemical.

48. Circle one answer. A homeowner is allowed to apply any pesticide, regardless of whether it is for general or restricted use.
    True   False

49. EO 12856 provides a pollution reduction goal for federal agencies of _____ percent by December 1999.

50. Excluding the general permit provisions, what can be dumped into the ocean under special permit?

51. Circle one answer. Federal agencies are required to reimburse the EPA for the cost of facility inspections under RCRA.
    True   False

52. Circle one answer. The Pollution Prevention Act requires that all government facilities file a Pollution Prevention Plan.
    True   False

53. Executive Order 12856 requires federal agencies to respond to the requirements of which federal laws?

54. Circle one answer. Material Safety Data Sheets provide environmental, health, and safety information on a commercial product. They are required to be available for employee reading whenever a hazardous chemical is used at a facility.
    True   False

55. What federal agency establishes tolerances for pesticide residues on raw and processed foods?

56. Circle one answer. A pesticide must be registered with the Department of Commerce before it can be sold in the market place.
    True   False

57. When an endangered species is likely to be harmed as a result of a federal action, the Endangered Species Act requires consultation with which agency?

58. Which federal agency protects marine mammals through the Marine Mammal Protection Act?

59. Circle one answer. A Finding of No Significant Impact (FONSI) may result from developing an Environmental Impact Statement.
    True   False

60. Name three differences between an Environmental Impact Statement and an Environmental Assessment.

61. Under what authority are NEPA documents required for major federal actions in foreign lands and waters?

62. What NEPA documentation does not require public involvement?

63. Circle one answer. Executive Order 12114 requires consideration of farmland protection in the NEPA process.
    True   False

64. The Oil Pollution Act of 1990 amended which statute?

65. Circle one answer. Primary drinking water standards are regulated to protect the health of the consumer; secondary drinking water standards are regulated to protect the public welfare.
    True   False

66. Circle one answer. Both primary drinking water standards and secondary drinking water standards are federally enforceable.
    True   False

12.2 ANSWERS

1. What important part of a *Federal Register* rule is not codified in the CFR? The preamble of a rule is not codified in the CFR. Only the regulation part, without rationale or explanation, is codified.

2. What national program does Section 303 of the Clean Water Act address? Section 303 of the Clean Water Act provides authority for the national water quality standards program.

3. What is the current OSHA PEL for asbestos in fiber per cc? The OSHA PEL for asbestos is 0.1 fiber per cc.

4. Name two of the four components of a Water Quality Standard. The four principal components of a water quality standard are:
   a. The designation of particular water uses for each waterway segment within the state.
   b. Numeric (or general) water quality criteria to protect the designated water uses.
   c. An antidegradation policy.
   d. An enforcement plan for the antidegradation policy.

5. For what does Section 312 of the Clean Water Act require that standards be promulgated? The discharge of sewage from a vessel and uniform national discharge standards.

6. When (approximately) was the first federal water pollution control legislation enacted? The first water pollution control legislation was the River and Harbor Act of 1899.

7. What is the purpose of water quality criteria in water quality standards? Water quality criteria protect the designated water uses of ambient waters. They provide the quality conditions to achieve those designated uses.

8. Circle the most correct answers. An NPDES permit is needed for:
   a. The discharge of dredge and fill material.
   (b.) Any discharge of toxic material into water of the United States.
   c. Any emission of ozone.
   (d.) Any pollutant discharge to waters of the United States.

9. Circle the most correct answer. Permits for dredge and fill material are issued by:
   a. Environmental Protection Agency.
   (b.) Corps of Engineers.
   c. Fish and Wildlife Service.
   d. Coast Guard.

10. Circle one answer. Federal regulations are promulgated by the Congress.
    True   (False)
    Applicable federal agencies promulgate regulations to interpret and implement laws passed by the Congress.

11. Circle one answer. Proposed rules are enforced by the EPA.
    True   (False)
    Proposed rules are unenforceable. They provide an indication of the policy of an agency. When developed and promulgated as an "interim rule" or "final rule," they have the force and effect of law.

12. Circle one answer. The District of Columbia area is in attainment for ozone.
    True   (False)

13. Circle one answer. The District of Columbia area is in non-attainment for carbon monoxide.
    (True)   False

14. What group of substances are regulated by the NESHAP for Ship Building and Ship Repair?   Volatile organic compounds (VOCs) in paint.

15. What is the required disposal method for liquids with PCBs equal to or greater than 500 ppm?   Incineration.

16. Circle one answer. PCBs can still be introduced into commerce.
    (True)   False
    But only in a totally enclosed manner.

17. If you circled "true" in question 16, what is the form in which PCBs can still be introduced into commerce?   Totally enclosed.

18. Circle one answer. TSCA banned certain uses of PCBs in 1976. However, the continued use of pre-1976 PCBs was grandfathered into the law.
    True   (False)
    TSCA does not provide for grandfathering PCB use. The law states that the continued use of PCB contaminated material, in a concentration that exceeds the regulated concentration, is a violation of law. An exception is provided when PCBs are totally enclosed.

19. Under TSCA, a manufacturer can use another manufacturer's environmental data for a product to meet permit application requirements, providing compensation is provided to the initial collector of the data.
    (True)   False

20. What is "friable" asbestos?   Asbestos whose fibers can easily enter the atmosphere or that can be crushed between thumb and forefinger.

21. What three control measures can be employed when severely damaged friable asbestos is found?   Mitigation measures may include operation and maintenance with surveillance every six months, repair, encapsulation, enclosure, and removal.

22. Under what section of TSCA can the EPA take action to prohibit or restrict the introduction into commerce of a chemical or chemical substance?   Section 6, the section under which asbestos and PCB bans were implemented.

23. Conducting a toxicity test, i.e., subjecting specific test organisms to a particular waste concentration, is often a requirement of an NPDES permit.
(True)   False
Whole effluent toxicity testing is increasingly becoming a condition of new and renewed NPDES permits.

24. Executive orders are signed by the President to require industry to stop polluting.
True   (False)
The President has direct control over federal agencies only. Executive orders are directed to federal agencies. The President has no direct control over industries. Industries are controlled through regulations that are promulgated in the public domain and may be challenged in the courts.

25. Circle all that are appropriate. In developing a draft permit, a permit writer may use:
(a.) Effluent guidelines.
(b.) Toxicity testing results.
c. Executive orders.
(d.) Best engineering judgment.
(e.) Water quality standards.

26. When is a solid waste a hazardous waste?   A solid waste is a hazardous waste when it is a listed hazardous waste, exhibits the characteristics of a hazardous waste, is a discarded commercial product, or is a hazardous waste in the judgment of the generator.

27. Circle one answer. An identification number is not needed to haul hazardous waste interstate?
True   (False)
An identification number is needed from the time of generation of a hazardous waste to its disposal.

28. Hazardous waste can be stored on-site without a permit for _____ 90 _____ days.

29. A small quantity generator is one who generates between _____ 100 _____ and _____ 1000 _____ kg/month of hazardous waste.

30. Circle one answer. Every permit issued for a discharge to water, of any kind whatsoever, must ensure that water quality standards shall not be violated.
(True)   False

31. Small generators of hazardous waste can legally store hazardous waste for _____ 180 _____ days.

32. Circle all that are appropriate: In the disposal of dredged spoil:
(a.) Clean Water Act Section 404(b) guidelines govern disposal in the ocean (out to 3 nm).

     b. The Corps of Engineers always has the last word on permit issuance.
     c. The Corps of Engineers may issue a permit for the transportation of
        dredged material for the purpose of dumping it into ocean waters.
     d. It is never necessary to conduct toxicity tests prior to disposal.

33. The CFR is another way of indexing federal laws.
    True   False
    The CFR codifies regulations, not laws.

34. The National Priority List is a list of (circle all that are appropriate):
     a. Chemicals that must be analyzed for in an NPDES permit application.
     b. Sites with threatened releases of hazardous substances.
     c. Fast food places along Interstate 70.
     d. Pesticides for intensive environmental study.

35. Circle one answer. The National Contingency Plan provides instructions
    for the clean up of hazardous waste sites only.
    True   False
    The National Contingency Plan addresses oil spills and hazardous waste
    sites.

36. Circle one answer. Ocean dumping is still a viable option for the disposal
    of sewage sludge from the New York area.
    True   False
    The Ocean Dumping Ban Act of 1988 prohibits the ocean disposal of
    sewage sludge and industrial wastes.

37. Circle one answer. It is legal for a Navy ship to discharge floating solid
    waste in the ocean beyond 25 nm from land.
    True   False
    When Special Areas come into effect for the Navy, the discharge of
    garbage other than food wastes will be prohibited therein.

38. Circle one answer. The Coastal Zone Management Act regulates land use
    in a 25-mi wide parcel of land along the coastline.
    True   False
    The coastal zone extends inland from the shorelines only to the extent
    necessary to control shorelands, the uses of which have a direct and sig-
    nificant impact on the coastal waters.

39. Circle one answer. EPCRA provides credits to industry for pollution pre-
    vention planning.
    True   False
    EPCRA provides no credits.

40. Circle one answer. FIFRA authorized the issuance of experimental use
    permits to enable a manufacturer to develop data necessary to register a
    pesticide.
    True   False

41. Check correct answer. The Federal Facilities Compliance Act provided that public vessels are _____ are not _____ XX _____ generators of hazardous waste.

42. Circle one answer. An ocean dumping general permit provides for human burial at sea.
    (True)   False

43. Check correct answer. Industrial wastes have been _____ XX _____ have not been _____ prohibited by an act of Congress from being dumped into the oceans.
    Prohibited as of December 31, 1991 by the Ocean Dumping Ban Act.

44. Circle one answer. It is now unlawful for vessels, other than public vessels owned or operated by the DOD, to discharge plastics into the oceans.
    (True)   False
    According to the Act to Prevent Pollution from Ships, the discharge of plastics into the oceans became unlawful after December 31, 1993, except for vessels operated by the Navy.

45. Circle the appropriate answers. Under the Endangered Species Act:
    (a.) To "harass" a listed species is unlawful.
    (b.) To "pursue" means the same as to "take."
    (c.) To "watch" closely a listed species is to "harass."
    (d.) An endangered species is one in "short" supply.

46. Circle one answer. Under EO 12856 and EPCRA, any facility that has to fill out an EPA Form R for the release of a chemical must continue to file the form annually regardless of the quantity of the chemical released.
    (True)   False

47. The threshold for reporting on Form R by a federal agency generally is _____ 10,000 _____ lb of a listed EPCRA chemical.

48. Circle one answer. A homeowner is allowed to apply any pesticide, regardless of whether it is for general or restricted use.
    True   (False)

49. EO 12856 provides a pollution reduction goal for federal agencies of _____ 50 _____ percent by December 1999.

50. Excluding the general permit provisions, what can be dumped into the ocean under special permit?   Dredged spoil that meets permit requirements. Processed fish wastes that meet permit conditions.

51. Circle one answer. Federal agencies are required to reimburse the EPA for the cost of facility inspections under RCRA.
    (True)   False

52. Circle one answer. The Pollution Prevention Act requires that all government facilities file a Pollution Prevention Plan.

True   (False)
Only those facilities that meet the threshold for filing Form R or EPCRA.

53. Executive Order 12856 requires federal agencies to respond to the requirements of which federal laws?   The Emergency Planning and Community Right-To-Know Act and the Pollution Prevention Act.

54. Circle one answer. Material Safety Data Sheets provide environmental, health, and safety information on a commercial product. They are required to be available for employee reading whenever a hazardous chemical is used at a facility.
    (True)   False

55. What federal agency establishes tolerances for pesticide residues on raw and processed foods?   The Environmental Protection Agency under authority of FIFRA.

56. Circle one answer. A pesticide must be registered with the Department of Commerce before it can be sold in the market place.
    True   (False)
    A pesticide must be registered with the Environmental Protection Agency, not the Department of Commerce.

57. When an endangered species is likely to be harmed as a result of a federal action, the Endangered Species Act requires consultation with which agency?   The Fish and Wildlife Service.

58. Which federal agency protects marine mammals through the Marine Mammal Protection Act?   National Marine Fisheries Service (NOAA) under the Secretary of Commerce.

59. Circle one answer. A Finding of No Significant Impact (FONSI) may result from developing an Environmental Impact Statement.
    True   (False)
    A FONSI may be the result of an Environmental Assessment.

60. Name three differences between an Environmental Impact Statement and an Environmental Assessment.
    EIS Record of Decision vs. EA FONSI.
    EIS Scoping Meeting requirement vs. EA No Scoping Meeting.
    EIS may require research to resolve issues vs. EA based upon available information.
    EIS generally 100 to 150 pages and with potential appendices vs. EA of 25 to 35 pages.

61. Under what authority are NEPA documents required for major federal actions in foreign lands and waters?   Executive Order 12114.

62. What NEPA documentation does not require public involvement?   Use of Categorical Exclusions.

63. Circle one answer. Executive Order 12114 requires consideration of farm-
    land protection in the NEPA process.
    True   (False)
    Executive Order 12114 addresses environmental effects abroad on major
    federal actions.

64. The Oil Pollution Act of 1990 amended which statute?   Section 311 of
    the Clean Water Act.

65. Circle one answer. Primary drinking water standards are regulated to pro-
    tect the health of the consumer; secondary drinking water standards are
    regulated to protect the public welfare.
    (True)   False

66. Circle one answer. Both primary drinking water standards and secondary
    drinking water standards are federally enforceable.
    True   (False)
    Secondary drinking water standards are not federally enforceable, but
    may be enforced by a state.

13 Safety and Precautions

The Occupational Safety and Health Act of 1970 requires virtually every private employer to furnish each employee a place of employment free from recognized hazards likely to cause death or serious physical harm. The Occupational Safety and Health Administration (OSHA) is the administrative agency for the Act. The National Institute for Occupational Safety and Health (NIOSH) is one of nine operating components of the Center for Disease Control, Atlanta, GA, established as an operating health agency within the Public Health Service. The Center for Disease Control is the federal agency charged with protecting the public health of the nation by providing leadership and direction in the prevention and control of diseases and other preventable conditions. NIOSH conducts research, provides technical assistance to OSHA, and recommends standards for OSHA adoption and promulgation. OSHA promulgates standards, conducts inspections, and provides for their enforcement. OSHA may promulgate standards that differ from the NIOSH recommendations, based upon a consideration of economic and other factors.

NIOSH evaluates all known medical, biological, engineering, chemical, trade, and other information relevant to a potential hazard. The purpose is to formulate recommendations on limits of exposure to potentially hazardous substances or conditions in the workplace, and to establish appropriate preventive measures designed to reduce or eliminate adverse health effects. These recommendations then are published and transmitted to OSHA and the Mine Safety and Health Administration for use in promulgating legal standards.

NIOSH publishes a "Pocket Guide to Chemical Hazards." The Guide addresses 677 separate chemicals and chemical substances. The following information is presented in 14 separate columns:

Chemical name and formula
Synonyms
Exposure limits
The level immediately dangerous to life or health
Physical description
Chemical and physical properties
Incompatibilities
Measurement methods
Personal protection and sanitation
Respirator selection
Route of health hazard
Symptoms
First aid
Target organs

The Eighth Edition of the National Toxicological Program's Annual Report on Carcinogens contains 198 entries of chemical and chemical substances of which 29 entries are listed as "Known to Be Human Carcinogens" and 169 are listed as "Reasonably Anticipated to Be Human Carcinogens." The document provides descriptive information for each listing under the headings of carcinogenicity, properties, uses, production, exposure, and regulations. The agencies participating on the working group for the International Agency for Research on Carcinogens (IARC) to keep this document up to date include: Agency for Toxic Substances and Disease Registry, Center for Disease Control/National Institute for Occupational Safety and Health, Consumer Products Safety Commission, Environmental Protection Agency, Food and Drug Administration, National Institutes of Health/National Cancer Institute, National Institutes of Health/National Institute of Environment Health Services, National Institutes of Health/National Library of Medicine, and U.S. Department of Labor/Occupational Safety and Health Administration.

Subpart Z of OSHA's 29 CFR 1910.1000 presents standards for chemical carcinogens and other hazardous materials. If required, the employer must provide respirators at no cost to the employee, and ensure that they are used when working with a specific chemical. Generally, there is a requirement for respirator use in areas where an employee is exposed to the permissible exposure limit (PEL).

When conditions exist at or above the action level for a chemical substance or condition, an annual employee training program and medical surveillance of employees must be instituted. The training program includes health effects, engineering controls and work practices, use and limitations of respirators and protective clothing, purpose and description of the medical surveillance program, and other relevant issues. The annual medical surveillance requirement includes a medical and work history, a complete physical examination, a chest roentgenogram, and a pulmonary function test. In addition the physician must provide any recommended limitations related to the use of personal protective equipment, including respirators. An employee's medical record shall be maintained for the duration of employment plus 30 years.

The OSHA hazard communication section at 29 CFR 1910.1200 has as its purpose to ensure that information concerning the hazards of the chemicals employees are exposed to in the workplace are made known to those employees. Each container of hazardous chemicals in the workplace must be labeled, tagged, or marked with the identity of the hazardous chemical and with appropriate hazard warnings. Chemical manufacturers and importers shall obtain or develop a material safety data sheet (MSDS) for each hazardous chemical produced or imported. Each employer shall have a MSDS for each hazardous chemical that they use. Each MSDS shall contain:

1. The identity used on the label or the chemical and common name of all health-hazard ingredients that comprise 1 percent or more of the composition.
2. Physical and chemical characteristics of the hazardous chemical, such as vapor pressure, and flash point.
3. The physical hazards of the chemical, including the potential for fire, explosion, and reactivity.

4. The health hazards, including signs and symptoms of exposures, and any medical conditions that are generally recognized as being aggravated by exposure to the chemical.
5. The primary route or routes of entry into the body.
6. The OSHA PEL, and any other exposure limit used or recommended by the chemical manufacturer, importer, or employer preparing the MSDS.
7. Whether the hazardous chemical has been found to be a carcinogen, and to what and by whom.
8. Any generally applicable control measures, such as appropriate engineering controls, work practices, or personal protective equipment.
9. Any generally applicable precautions for safe handling and use including appropriate hygienic practices, protective measures during repair and maintenance of contaminated equipment, and procedures for clean up of spills and leaks.
10. Emergency and first aid procedures.
11. The date of preparation of the MSDS.
12. The name, address, and telephone number of the responsible party preparing or distributing the MSDS.

14 Affected Environments

An aspiring employee of an environmental program should have a basic understanding of the environments that are protected by the environmental laws and regulations.

14.1 TEMPERATURE

Temperature is a prime regulator of natural processes within the water environment. It governs physiological functions in organisms and, acting directly or indirectly in combination with other water quality constituents, it affects aquatic life with each change. These effects include chemical reaction rates, enzymatic functions, molecular movements, molecular changes between membranes within and between the physiological systems, and organs of an animal. Because of the complex interactions involved, and often because of the lack of specific knowledge or facts, temperature effects as they pertain to an animal or plant are most efficiently assessed on the basis of net influence on the organism. Depending on the extent of environmental temperature change, organisms can be activated, depressed, restricted, or killed. Temperature determines those aquatic species that may be present; it controls spawning and the hatching of young, regulates their activity and stimulates or suppresses development; warmer water generally accelerates activity.

Temperature regulates molecular movement and thus largely determines the rate of metabolism and activity of all organisms, both those with a relatively constant body temperature and those whose body temperature is identical to, or follows closely, the environmental temperature. Because of its capacity to determine metabolic rate, temperature may be the most important single environmental entity to life and life processes.

Variations in temperature of streams, lakes, estuaries, and oceans are normal results of climatic and geologic phenomena. Waters that support some form of aquatic life other than bacteria or viruses range in temperature from, 26.6°F in polar sea waters to 185°F in thermal springs. Most aquatic organisms tolerate only those temperature changes that occur within a narrow range to which they are adapted, whether it is high, intermediate, or low on the temperature scale.

Pure water reaches its maximum density or weight, at 39.2°F (4°C). Water becomes lighter as it cools or warms from its maximum density. Water pressure increases with depth. At a depth of 100 ft the water pressure is 58 lb/in.^{2} or approximately 4 times the pressure at the surface.

The seasons induce a cycle of physical and chemical changes in the water that are often conditioned by temperature. In a lake, for a few weeks in the spring, water temperatures may be homogeneous from the top of a water body to the bottom. Vertical water density is also homogenous and it becomes possible for the wind to mix the water, distributing nutrients and flocculent bottom solids from the deeper waters to the very surface. Oxygen is mixed throughout the water during this time. The advance of summer quickly checks circulation by warming the surface waters; as they warm they become lighter, resting over colder water of greater density.

Thus, a permanent thermal stratification is formed for many months. In natural deep bodies of water three layers form. The upper layer, or epilimnion, represents the warm, more or less freely circulating region of approximately uniform temperature. This may vary in thickness from 10 ft or less in shallow lakes to 40 ft or more in deeper ones. The middle layer, or thermocline, is the region of rapid water temperature change, which has a decreasing temperature change of about 1.8°F for each 3.28 ft (1.0°C for each meter) in depth. The lower (profundal) region, or hypolimnion, is the cold region of approximately uniform temperature. It is isolated from circulation with the upper waters and receives no oxygen from the atmosphere during stratification.

As autumn comes, the surface of the lake cools; the epilimnion increases in thickness until the lake becomes homothermous, and again a period of complete circulation begins. This occurs from late September to December, depending upon the area and depth of the lake and its geographic and climatic location. This condition lasts until changes in density reestablish stratification, or until the lake is frozen over. Circulation ceases until spring.

Thermal stratification in reservoirs may assume many patterns depending on geographical location, climatological conditions, depth, surface area, type of dam structure, penstock location, and use for power. Reservoirs or impoundments are of two basic types: main stream and storage.

The main stream (run of the river) reservoir is typically an impoundment formed by a relatively low dam that rarely exceeds 60 to 80 ft in height. Much of the impounded water is restricted to the original channel, and water retention ranges from a few days to a few weeks. Man-regulated fluctuations in surface levels usually are controlled within a range of 2 to 3 ft. Main stream impoundments are used principally for navigation and for power production. Thermal stratification often consists of small but fairly regular gradients of 5°F to 10°F from top to bottom during the summer. This gradient is most likely to occur in a reservoir with limited surface area where wind action is moderate and velocities are low. Temporary thermoclines have been recorded where the temperature gradient is steep through a narrow band of water.

Another form of thermal stratification in main stream reservoirs involves the inflow of a stream of water that is colder than the normal surface water. Since the penstock intake (discharge) may extend from near the bottom to within 15 or 20 ft of the water surface, the cold stream of water flows through the impoundment, creating a thermocline below the water surface at the dam and extending upstream parallel to the bottom of the reservoir.

The storage reservoir, as its name implies, is used to impound water when surface runoff is high (i.e., flood flows) for release when runoff is low. As a result, the surface water level varies over a wide range, sometimes 70 ft or more during the year, and is generally highest at or near the end of a rainy season and lowest just before the next rainy season. The drawdown of the reservoir requires that the discharge intake be located deep in the reservoir, below the minimum level to which the water will be drawn.

The storage reservoir is often located at the headwaters of a stream that frequently has a steep slope. The dam is high, often more than 100 ft. The stored water spreads far beyond the former river channel into numerous fingers or embayments to provide a large surface area. Vertical cross sections of the reservoir are large in relation to stream flow, and flow velocities in the reservoirs are negligible. Water may be retained in the reservoir for many months. Passage of water through the reservoir may be discontinuous, and significant portions of the water may remain in storage for nearly a year.

Most storage reservoirs exhibit the classical type of thermal stratification described for natural deep-water bodies. Reservoirs that do not store substantial volumes of water at winter temperatures or that discharge such water before warm weather occurs do not develop thermoclines; neither do shallow reservoirs with broad expanses of surface areas exposed to strong winds that mix the waters. In the southern portion of the country where surface water temperatures rarely drop below 4°C there is no stratification in winter and temperatures are nearly uniform throughout the impoundment.

Density currents in reservoirs can occur when there are differences in temperature, differences in the concentration of electrolytes especially carbonates, and differences in the silt content. Four well-defined horizontal zones with respect to dissolved oxygen during thermal stratification have been noted in reservoirs. These include a well-aerated surface stratum, a zone of stagnant water within the thermocline, a second stratum of water rich in dissolved oxygen below the thermocline, and a bottom layer of stagnant water. In some instances, density currents have been detected from 60 to 80 ft below the surface. Density currents affect the fish population because game fish orient themselves both to the stratum of stagnant water caused by density currents and to the temperature range that suits them best. Often they become trapped by a lack of oxygen within this zone.

In the deep, stagnant, summer bottom waters, as well as in ice covered waters, atmospheric reaeration is absent and oxygen from photosynthesis by plants is limited. Decomposing organisms (especially those settling to the bottom waters in summer) remove oxygen from the water and the gaseous by-products of decomposition are trapped. Undesirable soluble phosphorus, carbon dioxide, iron, and manganese concentrations increase in these stagnant waters. Designed thermal discharges can reduce some of these problems. Ice cover can be limited, thus allowing wind and thermally induced currents to reduce winter stagnation. A deep-water summer discharge could warm hypolimnetic waters to decrease density and allow total water mass mixing where a cold water fishery would not be damaged by such action.

Fish and other aquatic life occurring naturally in each body of water are those that have become adapted to the temperature conditions existing there. The interrelationship of species, length of daylight, and water temperature are so interrelated that even a small change in temperature may have far-reaching effects. An insect nymph in an artificially warmed stream, e.g., might emerge from its mating flight too early in the spring and be immobilized by the cold air temperature, or a fish might hatch too early in the spring to find its natural food organisms because the food chain depends ultimately upon plants, and those in turn, upon the length of daylight, as well as temperature. The inhabitants of a water body that seldom becomes warmer than 70°F are placed under stress, if not killed outright, by 90°F water. Even at 75° to 80°F, they may be unable to compete ꜱuccessfully with organisms for which the higher temperature is favorable. Similarly, the inhabitants of warmer waters are at a competitive disadvantage in cool water.

Reproduction cycles may be changed significantly by increased temperature because this function takes place under restricted temperature ranges. Spawning may not occur at all because temperatures are too high. Thus, a fish population may exist in a heated area only by continued immigration. Disregarding the decreased reproductive potential, water temperatures need not reach lethal levels to wipe out a species. Temperatures that favor competitors, predators, parasites, and disease can destroy a species at levels far below those that are lethal.

Fish food organisms are altered severely when temperatures approach or exceed 90°F. Predominant algal species change, primary production is decreased, and bottom associated organisms may be depleted or altered drastically in numbers and distribution. Increased water temperatures may cause aquatic plant nuisances when other environmental factors are favorable.

14.2   DISSOLVED OXYGEN

Dissolved oxygen (D.O.) in appropriate concentrations is essential to keep organisms alive, and to sustain the species reproduction, vigor, and the development of populations. Organisms undergo stress at reduced D.O. concentrations that make them less competitive to sustain their species within the aquatic environment. For example, D.O. concentrations around 3 mg/l or less have been shown to interfere with fish populations through delayed hatching of eggs, reduced sizes and vigor of embryos, production of monstrosities in young, interference with food digestion, acceleration of blood clotting, decreased tolerance to certain toxicants, reduced food efficiency and growth rates, and reduced maximum sustained swimming speed.

Oxygen enters the water by absorption directly from the atmosphere or by plant photosynthesis, and is removed by respiration of organisms and by decomposition of organic matter. Oxygen derived from the atmosphere may be by direct diffusion or by surface water agitation by wind and waves, which may also release dissolved oxygen under conditions of supersaturation.

In photosynthesis, aquatic plants use carbon dioxide and liberate dissolved and free-gaseous oxygen at times of supersaturation. Since energy is required in the form of light, photosynthesis is limited to the photic zone where light is sufficient to facilitate this process. During respiration and decomposition, animals and plants consume dissolved oxygen and liberate carbon dioxide at all depths where they occur. Because excreted and secreted products and dead animals and plants sink, most of the decomposition takes place in the hypolimnion in lakes; thus, during lake stratification there is a gradual decrease of dissolved oxygen in this zone. After the dissolved oxygen is depleted, anaerobic decomposition continues with the evolution of methane and hydrogen sulfide gases.

In the epilimnion, during thermal stratification, dissolved oxygen is usually abundant and is supplied by atmospheric aeration and photosynthesis. Phytoplankton are plentiful in fertile lakes and are responsible for most of the photosynthetic oxygen. The thermocline is a transition zone from the standpoint of dissolved oxygen, as well as temperature. The water rapidly cools in this region, incident light is much reduced, and photosynthesis is usually decreased. If sufficient dissolved oxygen is present, some cold water fish abound. As dead organisms that sink into the hypolimnion decompose, oxygen is used. Consequently, the hypolimnion in fertile lakes may become devoid of dissolved oxygen following a spring overturn, and this zone may be unavailable to fish and most benthic invertebrates at such time. During the two brief periods in the spring and fall when lake water circulates, temperature and dissolved oxygen are the same from top to bottom and fish can use the entire water depth.

14.3   pH

All indications are that pH values above 5.0 and ranging upward to pH 9.0 are not lethal to most fully developed freshwater fishes. Much more extreme pH values, below 4.0 and above 10.0 can be tolerated by resistant species. However, such extreme pH values are undesirable and hazardous for fish life in waters which are not naturally so acid or alkaline. Fish mortalities occur below a pH of 5.0.

14.4   LIGHT

Rooted, suspended, and floating aquatic plants require light for photosynthesis. The principal factors affecting the depth of light penetration in natural waters include suspended microscopic plants and animals, suspended mineral particles such as mineral silt, stains that impart a color, detergent foams, dense mats of floating and suspended debris, or a combination of these. The region in which light intensity is adequate for photosynthesis is often referred to as the trophogenic zone, the layer that encompasses 99 percent of the incident light. Generally, 1 percent of the incident water surface light is necessary for photosynthesis in algae, and perhaps 2.5 percent of the incident light is necessary for photosynthesis in the vascular aquatic plants. The depth of the trophogenic zone may vary from less than 5 ft in eutrophic lakes to greater than 90 ft in oligotrophic lakes. Thus, light can become a controlling factor for biotic production in many water bodies. Especially, light may restrict the growth of rooted aquatic plants to the littoral region in many lakes.

14.5   FLOW

The velocity of water movement is extremely important to aquatic organisms in a number of ways including the transport of nutrients and food past those organisms attached to stationary surfaces; the transport of plankton and benthos as drift, which in turn serve as food for higher organisms; and the addition of oxygen to the water through surface aeration. Silts are moved downstream and sediments may be transported as bed load. These, in turn, may be associated with major nutrients such as nitrogen and phosphorus, which may be released at some point downstream from their introduction.

Flow determines those species of stream bed organisms that may be present in a particular stream reach. Black fly larvae, e.g., require fast moving water. Immature forms of caddisfly and mayfly will develop to large populations in swiftly moving, but not fast and turbulent water. Among many invertebrate genera there are those particular species that are adapted for life not only under the two extremes of flow but also under its many variations.

14.6   MAJOR NUTRIENTS

Eutrophication is a term meaning enrichment of waters by nutrients through either person-created or natural means. Present knowledge indicates that the fertilizing elements most responsible for lake eutrophication are phosphorus and nitrogen. Iron and certain trace elements are also important. Sewage and sewage effluents contain a generous amount of those nutrients necessary for algal development.

Lake eutrophication results in an increase in algal and weed nuisances and an increase in midge larvae, whose adult stage has been a nuisance for humans around many lakes. Dense algal growths form surface water scums and algal-littered beaches. Water may become foul-smelling. Filter-clogging problems may occur at municipal water installations. When algal cells die, oxygen is used in decomposition, and fish kills have resulted. Rapid decomposition of dense algal scums, with associated organisms and debris, gives rise to odors and hydrogen sulfide gas that creates strong citizen disapproval; the gas has often stained the white lead paint on residences adjacent to the shore.

Nitrogen and phosphorus are necessary for life in water. They are also necessary components of an environment in which excessive aquatic growths occur. Algal growth is influenced by many varied factors; vitamins, trace metals, hormones, auxins, extracellular metabolites, autointoxicants, viruses, and predation and grazing by aquatic animals. Several vitamins in small quantities are requisite to growth in certain species of algae. In the freshwater environment, algal requirements are met by vitamins supplied in soil runoff, lake and stream bed sediments, solutes in the water, and metabolites produced by actinomycetes, fungi, bacteria, and several algae.

Evidence indicates that:

1. High phosphorus concentrations are associated with accelerated eutrophication of waters, when other growth promoting factors are present.
2. Aquatic plant problems develop in reservoirs or other standing waters at phosphorus values lower than those critical in flowing streams.
3. Reservoirs and other standing waters collect phosphates from influent streams and store a portion of these within consolidated sediments.
4. Phosphorus concentrations critical to noxious plant vary, and they produce growths in one geographical area, but not in another.

Generally, phosphorus is considered to be the limiting nutrient to aquatic production.

Once nutrients are combined within the ecosystem of the receiving water, their removal is tedious and expensive. Removal must be compared to quantities in inflowing waters to evaluate accomplishment. In a lake, reservoir, or pond, phosphorus is removed naturally only by outflow, by insects that hatch and fly out of the drainage basin, by harvesting a crop, such as fish, and by combination with consolidated benthic sediments. Even should adequate harvesting methods be available, the expected standing crop of algae per acre exceeds 2 tons and contains only about 1.5 lb of phosphorus. Similarly, submerged aquatic plants could approach at least 7 tons per acre (wet weight) and contain 3.2 lb per acre of phosphorus. Probably only half of the standing crop of submerged aquatic plants can be considered harvestable. The harvestable fish population (500 lb) from 3 acres of water would contain only 1 lb of phosphorus.

14.7   MINOR NUTRIENTS

Minor nutrients or micronutrients include iron, manganese, copper, zinc, molybdenum, vanadium, boron, chlorine, cobalt, and silicon. These can become key elements in photosynthesis and to the general health of the aquatic system. Such micronutrients are essential for growth and development, but when present in abundance they are toxic to aquatic systems.

14.8   POLLUTANTS

The effects of pollution upon the water environment assumes many characteristics, as well as an infinite variation in degree. The specific environmental and ecological responses to a given pollutant depends largely on its volume combined with the characteristics of the wastewater, and the volume and characteristics of the receiving water into which it flows. Pollutants may provide an aesthetic insult; a toxic action; a blanketing effect that destroys the stream or lake bed; a biodegradable, organically-decomposable material that removes the dissolved oxygen from the water; a hazard to the health of man and other animals that use the water; a substance that magnifies in concentration as it becomes escalated through the aquatic food web; an alterant of water temperature which is the prime regulator of natural processes within the water environment; and a supplier of fertilizing nutrients that stimulate excessive production among some aquatic species.

The discarded objects, especially in streams, that result in an aesthetic environmental insult can be an asset as far as life in the water is concerned. In some situations, the discarded automobile tire or baby carriage may provide additional substrate area for some types of organisms to colonize, provided other aspects of the physical and chemical characteristics of the waterway support such biotic development. Many streams naturally lack appropriate substrate for the attachment of various macroorganisms that can supply food for a fishery. Artificial materials are often introduced in lakes and near oceanic beaches as a management practice both to furnish a site for the development of fish food organisms and to act as a harbor of security for the fish population that is attracted to the device. In these cases, notably rare, the aesthetic liability of the discarded material must be weighed against the assets for the production of a higher use waterway.

14.8.1   INORGANIC SILTS

The general effect on the aquatic environment of inorganic silts is to reduce severely both the kinds of organisms present and their populations. As particulate matter settles to the bottom it can blanket the substrate and form undesirable physical environments for organisms that would normally occupy such a habitat. Erosion silts alter aquatic environments directly by screening out light, by changing heat radiation, by blanketing the stream bottom and destroying living spaces, and by retaining organic matter and other substances that can create unfavorable conditions. Developing eggs of fish and other organisms may be smothered by deposits of silt; fish feeding may be hampered. Direct injury to fully developed fish, however, by non-toxic suspended matter occurs only when concentrations are higher than those commonly found in natural water or water associated with pollution.

14.8.2   TOXIC MATERIALS

Wastes containing concentrations of heavy metals or other toxic substances either individually or in combination may be toxic to aquatic organisms and thus have a severe impact on the water community. Fish kills are often the result of direct toxicity. Such acute toxicity may be so broadly effective that many life forms are damaged at one time, or it may be highly selective. Acute toxicity may result from a low concentration of a highly toxic material or a high concentration of a less toxic material.

Toxic actions that may require weeks or months to be noticed is referred to as low-level, cumulative, or chronic toxicity and is most often observed as a reduction in the production of a particular type of organism. Slowly toxic materials may be more deleterious to a particular developmental stage rather than to the adult organism. Low-level toxicities may change the entire population balance in a particular ecosystem. There are a number of processes that may take place:

1. Susceptible species of either fish or fish food organisms may gradually die off thereby permitting tolerant species that are less desirable to man to flourish because of a lack of competition.
2. If algae or invertebrate food organisms are killed by a low-level toxicity, fish may die or move out of the area because of an inadequate food supply.
3. Weakened individuals surviving near the threshold of their tolerance are more susceptible to attack by parasites and disease such as the aquatic fungus, *Saprolegnia.*
4. Reproduction potential may be altered because eggs or very young individuals may be more susceptible to the low-level toxic substance than are the adults. The end result can be a slow and subtle alteration of the characteristics of the stream or lake biological community.

14.8.3   ORGANIC POLLUTION

What happens to water inhabiting organisms when organic wastes are introduced? Organic or biodegradable wastes are attacked by bacteria upon entering the water environment and during this process of decomposition the dissolved oxygen, so necessary for life within water, is used and reduced. When the organic load to the receiving waterway is heavy, all of the dissolved oxygen may be used and the waterway becomes anaerobic and, from the standpoint of aquatic life, virtually dead. In addition to the oxygen-consuming properties associated with organic wastes, solids may settle to the waterway's bottom forming sludge banks that not only continue to exert an oxygen demand during the decomposition process, but also furnish an ideal habitat for the development and reproduction of tremendous numbers of sludgeworms and associated undesirable organisms.

Upstream from the introduction of organic wastes, classic description details a clean-water zone or one that is not affected by pollutants. A clean-water zone tends to have a large variety of fish and bottom-associated organisms. Predators within the biotic community keep the system in balance. Many species but relatively few organisms for a given species is the general rule. The water is clear and silt and sludge free. Algae in the flowing water is the basis of the aquatic food web. At the point of waste discharge, and for a short distance downstream, there is formed a zone of degradation where wastes become mixed with the receiving waters, and where the initial attack is made on the wastes by bacteria and other organisms in the process of decomposition.

Following the zone of degradation, there is a zone of active decomposition that may extend for miles, or days of stream flow, depending in large measure on the volume of dilution that is afforded the waste by the stream and the water temperature that affects the rate of biological decomposition. The biological process that occur within this zone are similar in many respects to those that occur in a wastewater treatment plant utilizing a biological process of waste treatment. Within this zone, waste products are decomposed and those products that are not settled as sludge are assimilated by organisms in life processes.

Within the zone of active decomposition, conditions of existence for aquatic life are at their worst. The breakdown of organic products by bacteria may have consumed available dissolved oxygen. Sludge deposits may have covered the stream bed thus eliminating dwelling areas for the majority of bottom-associated organisms that could be found on a similarly unaffected area. Fish spawning areas have been eliminated, but perhaps fish are no longer present because of the diminished oxygen supply and the substantially reduced available natural food. Here, aquatic plants will not be found in large numbers because they cannot survive on the soft, shifting blanket of sludge. Turbidity may be high and floating plants and animals destroyed. Water color may be substantially changed. When organic materials are decomposed, a seemingly inexhaustible food supply is liberated for those particular organisms that are adapted to use this food source. Thus, bacteria and certain protozoan populations may increase to extremely high levels. Those bottom-associated organisms such as sludge worms, bloodworms, and other worm-like animals may also increase to tremendous numbers because they are adapted to burrowing within the sludge, deriving their food therefrom, and existing on sources and amounts of oxygen that may be essentially non-detectable by conventional field investigative methods. The general rule is that the zone of active decomposition is inhabited by a relatively few species, but there are great numbers of individuals representing some of the species. For example, it is possible to find in excess of 50,000 sludgeworms living within each square foot of bottom area under conditions of severe organic pollution.

A zone of recovery follows the zone of active decomposition. The recovery zone is essentially a reach of water in which the quality is gradually returned to that which existed prior to the entrance of pollution. Water quality recovery is accomplished through physical, chemical, and biological interactions with the aquatic environment.

Finally, the zone of recovery terminates in another zone of clean water or an area unaffected by pollutants that is similar in physical, chemical, and biological features to that which existed upstream from the pollution source.

14.9   WATER QUALITY INVESTIGATION

14.9.1   OBJECTIVES

Study objectives are a necessary and important beginning to any investigation. Careful thought and consideration should be given to their development. The objectives should encompass clear, concise, positive definitions of the investigation's purpose, its scope, and its boundary limits. Study objectives should be realistically oriented to the number, completeness and disciplines of the investigation. Objectives should be adjusted to the budgetary limitations for the study, as well as to the length of time allocated for the study, including the final report preparation. Ultimately, as the study progresses, and at its conclusion, the study's success and accomplishments will be judged in part on the extent to which it fulfilled the objectives stated at the study's instigation. Study objectives become important tools to guide subsequent investigations and to delineate avenues of approach toward problem solving.

Study objectives should be committed to writing as the first act in formulating an essential study plan. When properly developed, they will ensure adherence to the essential investigation and discourage pursuit of the interesting but nonessential bypaths or tangential considerations that so often dominate and defeat a well-intended purpose. Written objectives fix the responsibility of those charged with supervision of the study and they provide a basis for judging the extent to which the results meet the needs that justified the initial undertaking.

Objectives of water quality studies that require samples from a single location or isolated locations would include:

- The establishment of a baseline record of water quality.
- Investigation of the suitability of an area for a water supply source for municipal, industrial, agricultural, recreational, or other use.
- Monitoring wastewater discharges and their effects at a particular site.
- Research or demonstration on analytical procedures for water quality examination.

Objectives that require the examination of water quality from locations at related points on a waterway include:

- Determination of the nature and extend of pollution from point or non-point sources.

- Determination of adherence to or violation of water quality standards.
- Determination of characteristics and rates of natural purification of waterways.
- Determination of causes of fish kills and other catastrophic events involving water quality deterioration.
- Determination of existing water quality through a waterway reach prior to some anticipated event that is expected to alter water quality.
- Research and demonstration of techniques of waterway investigation.
- Serve as an environmental training laboratory to train water quality investigators.
- Serve as a basis for predicting water quality changes that may be caused by anticipated increased pollution loads or the implementation of certain pollution abatement or control activities.
- Serve as a data base for projecting a cost-benefit analysis of a water management effort.

14.9.2 PLANNING

Planning for a waterway investigation involves a myriad of details that are essential for the completion of a successful study. The first essential activity is to become familiar with available information on the waterway under investigation that may relate to the present activity. Seldom is the investigation an original event. Most lakes and streams in the nation have been investigated to some extent by someone and many prognostications and predictions have been recorded by previous investigators relating to the quality of water within a given reach of the waterway or the effects of significant anticipated changes in the watercourse that may affect water quality. The results of these investigations have been recorded in either published or unpublished reports, the latter of which may reside in a now obscure file of an appropriate state agency. Much time and effort in redoing what already has been done can be saved on the part of the investigator by searching out and becoming familiar with the past studies that relate to the watercourse in question.

Good field maps of the watercourse under investigation must be secured and the points of access noted. Factors affecting the investigation should be recorded on the map and these include locations of various water uses, geographical boundaries such as state limits of other significant landmarks, and marked changes in water characteristics such as the entry of free-flowing streams into reservoirs or lakes. The approximate location of known waste sources such as industries, municipalities, or other significant waste contributors should be marked. The approximate length of the waterway to be investigated should be noted, as well as the area of the watercourse that will impact the study. Tentative sampling locations should be selected from the maps based on points of access and stream mile designations developed for major landmarks on the waterway and the location of waste sources. Development of these conditions necessitates that the selected maps be accurate and of suitable scale.

Following a desk top analysis of available background data and a perusal of information gathered on the waterway by previous investigators, a reconnaissance survey should be undertaken whenever time permits. During a reconnaissance survey a judgment can be reached on the potential effects on water quality of individual and combined waste sources, the reach or reaches of the waterway that are of potentially greatest concern in the particular investigation, and points of access and anticipated sampling locations. Certain judgments should be reached during the reconnaissance survey that will save much time and effort at a later date. These would include the advantages and disadvantages of sampling by boat as opposed to collecting samples from bridges and by wading, or by a car top or trailered boat that may be lowered into the water from several points of access along the waterway. Reaching this decision will necessitate observations on stream width, depth, nature and types of stream substrate, relative flow, as well as any morphometric features that would influence a sampling procedure. The availability of boats for rental along the waterway and the availability of suitable access points for the types of samples to be collected should be ascertained during such reconnaissance. Also, contacts may be made with local officials or local investigators who may be encouraged to participate in some manner with the actual investigation. Arrangement should be made with landowners to cross private lands at times when samples are to be collected from the waterway should this be a necessity.

A few samples should be collected from readily available access points along the waterway during the reconnaissance survey to ascertain the relative water quality at various points and to aid in the judgment of selecting sampling locations. This may involve the collection of water samples or certain chemical analyses, and it should involve the examination of rocks, twigs, or submerged debris to ascertain the types of biota that are able to exist in a given reach of the waterway. Much can be determined about the water quality through a cursory examination of the attached organisms that are found on the top and bottom of submerged rocks and twigs. Observations of visible conditions along a waterway associated with a brief examination of attached organisms on submersed objects should be sufficient to delineate appropriate sampling locations for future investigation.

Following the completion of a reconnaissance survey, and subject to modification during the course of actual field sampling, decisions can be made on:

1. Types of samples necessary to meet the objectives of the study (i.e., various physical, chemical, and biological samples).
2. Sampling locations for each of the selected types of samples.
3. Periodicity of sampling and approximate time necessary for the collection of a specific sample.
4. Approximate numbers of samples necessary to meet the objectives of the study.

The next aspect of study planning involves the details necessary to initiate the process of data collection. Decisions must be made on methods of sample handling between the point of collection and the point of analytical result, sample preservation, and transportation of samples to a base laboratory. Often biological samples may be preserved for examination at some future and more convenient time. Certain samples for chemical analyses also may be preserved for a short time until they can be analyzed at an appropriate laboratory where precision instruments are available. Sample collection containers must be obtained and the number of these will depend upon the relative number of samples to be collected during the investigation. Sampling equipment, data cards, notebooks, and all of the necessary paraphernalia associated with the collection, retention, and shipment of samples must be obtained and organized. Collection and field analytical gear should be checked and rechecked to determine that no essential piece of equipment has been omitted and to determine that the equipment is operable and functions according to designed specifications.

If a study is some distance from the base of operations, a portion of the study planning involves the making of travel arrangements, room accomodations, transportation of samples and equipment to and from the sampling areas, and arrangements for such items as transportation during the investigation, procurement of outboard motor gasoline, cartons or boxes for shipping collected samples, ice to keep certain samples cold if this is a prerequisite for analyses, and other considerations. Laboratories that will analyze collected samples should be alerted and given an estimate of the number and kinds of samples that will be submitted, the types of analyses required, how the samples will be shipped, and the approximate dates of arrival. The laboratory should also be advised of the date their analytical results will be required to meet the investigation's deadline.

A preliminary cost estimate can now be made of the investigation under consideration and it may be that the first compromise of the ideal plan will be necessary. The cost must be adjusted to the available budget. The compromise may be in a reduction in the number of sampling locations, in the types of analyses to be made, in the number of samples to be obtained from each location, or any combination of these factors. Realistically, the conceived ideal for a field investigation is seldom achieved because of resource limitation or an alteration in the time schedule that must be imposed upon the study.

There are a number of sources of information concerning investigations connected with water quality in waterways. One of the best sources is the state water pollution control agency. As would be expected, this agency usually has the most complete collection of information and data on factors involved with water quality within a state. They have conducted surveys over a number of years and have received complaints and statements from citizens concerned with this subject. Other state agencies that may have important data include the State Health Department, which is generally responsible for supervising public water supplies; the State Fish and Game Department; the State Geological Survey, which cooperates with the U.S. Geological Survey in the stream gaging program; and the Public Service Commission, which usually has jurisdiction over dams and obstruction to navigation. In addition, interstate agencies such as the Interstate Commission on the Potomac River Basin, the Delaware River Basin Commission, and the Ohio River Valley Water Sanitation Commission usually have information similar to that in State Water Pollution Control Agency files. The U.S. Environmental Protection Agency with its ten regional offices located in strategic cities throughout the United States has a great deal of particular information on the waterways of the nation. The regional office in question should be contacted for water quality information as it may pertain to a particular study. Federal river development agencies, such as the U.S. Corps of Engineers, the Bureau of Reclamation of the U.S. Department of the Interior, and the Tennessee Valley Authority are all fertile sources of information on streams for which they have responsibility. The U.S. Geological Survey operates stream gaging stations and reports daily stream discharge information throughout the nation, usually in cooperation with the states. The U.S. Bureau of Sport Fisheries and Wildlife, and the U.S. Marine Fisheries Service collect data on fish and fishing. Municipal water treatment plant operators have comprehensive records on the quality of the water serving their source of raw water. Often these data include both chemical and biological constituents and the record of time for which these data have been collected extends through many years. Wastewater treatment plant operators often keep a log of the quality of their wastewater and, in rare instances, have data on stream water quality both upstream and downstream from their discharge point.

14.9.3 DATA COLLECTION

On approaching a stream sampling location, a number of observations are recorded. These include water depth; presence or riffles and pools; width; flow characteristics; bank cover; presence of slime growths, attached algae, floating algae, and other aquatic plants; unusual coloration of the water and stream bed; and unusual physical characteristics such as silt deposits, organic sludge deposits, iron precipitates, or various waste materials from industrial processes.

Organisms associated with the stream bed are valuable to relate to water quality because they are not equipped to move great distances through their own efforts and, thus, they remain at fixed points as indicators of water quality. Because the life history of many of these organisms extends through one year or longer, their presence or absence is indicative of water quality within the past, as well as the present. Bottom-associated organisms are relatively easy to capture with conventional sampling equipment and the amount of time and effort devoted to their capture and interpretation is not as great as that required for other components of the aquatic community.

A qualitative determination of bottom-associated organisms often provides information sufficient to determine relative water quality. The qualitative search for benthos should involve the collection of organisms from rocks, plants, submersed twigs or debris, or leaves of overhanging trees that become submersed and waterlogged. It is often convenient to scrape and wash organisms from these materials into a bucket or tub partially filled with water and then to pass this water through a U.S. Standard Series No. 30 soil sieve to concentrate and retain the organisms. The collected sample may be preserved for organism sorting and identification later. The investigator should search until the majority of species have been collected.

The basic principle in qualitative sampling is to collect as many different kinds of organisms as practical. Two convenient limiting methods are:

1. Presetting a time limit for collection at each sampling location to 30 min or 1 h.
2. Sampling in an area until new forms are encountered so infrequently that the law of diminishing returns dictates abandoning the sampling location.

It is important to ascertain that the sampling location is representative of stream conditions and that the sample collected is representative of and contains those forms predominant in the area. To make comparisons among upstream and downstream sampling locations, it is important that the physical conditions of the respective sampling sites are similar insofar as possible. The physical conditions of the sampling site should be recorded in the sampling notebook for later use in making comparisons among sampling locations.

Commercial sampling devices are available that are designed to capture organisms within a particular area. If population comparisons are being made among sampling locations upstream and downstream from a particular area, it is important to use the same type of sampling device and strategy at all sampling locations to ensure comparability of sampling data. Sampling devices that may be used include various types of dredges, drift nets to capture drifting organisms, artificial substrates, stream riffle samplers, and similar devices.

14.9.4 LAKES

A complex interaction of many physical, chemical, and biological factors, influenced by meteorological phenomena, occur in a lake environment. Nutrients are vital as a basic food supply and dissolved oxygen is an essential component for aquatic life. Changes in the concentrations of either of these factors will induce significant alterations in the complexity of life that inhabits lake waters. The pond or lake environment tends to be a vertical environment as contrasted with the more horizontal environment of the flowing stream. Because of this, factors such as temperature and light penetration assume roles of paramount importance and become controlling conditions to life in the vertical environment.

The standing water environment receives its inflowing water from tributary streams, generally land runoff, seepage from adjacent areas, and springs. The standing water basin is, thus, the first line receptacle for those materials that may be washed or drained from the lands within the drainage basin. Large lakes are directly impacted by pollutants discharged from the atmosphere as well. Those materials that are discharged from the pond, lake, or other standing water bodies into downstream waterways eventually reach the estuaries or the oceanic environments, which become the ultimate receptacle for such materials.

The discharge from a lake or reservoir also has an effect on downstream water quality. Water flowing from a natural lake should be expected to be of a quality similar to that of the water in the uppermost stratum of the lake. However, when water in a free-flowing stream is impounded in a large storage reservoir, marked changes are produced in the physical, chemical, and mineral quality of the water. In reservoirs operated for flood control and power production, discharge releases are often reduced over weekends and other periods of off-peak power loads. Discharges likewise may be increased substantially during the week days. The penstock usually is located deep within the reservoir and the temperature of the water discharged to the receiving stream may be substantially lower than the natural receiving water temperature. Discharged water also may be low in dissolved oxygen concentration and release odors of hydrogen sulfide from decaying organic materials in the deeper portions of the reservoir.

The ecology of the receiving stream is drastically altered as a result of a low-level discharge characterized by low temperatures and reduced dissolved oxygen concentrations. The warm water fish habitat may be destroyed. A cold water fish habitat may be created providing dissolved oxygen is sufficient. Bottom fauna may be changed in the receiving waterway from an assemblage of stoneflies and hell-grammites to an assortment of cold water species such as immature midges, black flies, caddisflies, and scuds. Such a substantial alteration of the physical and chemical characteristics of the receiving stream can only produce a drastic change in the composition of the biota inhabiting the receiving water environment.

A lake or reservoir is a settling basin for the water that it receives. It retains a portion of those nutrients that enter in its consolidated sediments and in the biota that it supports. The amount or percentage retained is dependent upon:

1. The nutrient loading to the lake or reservoir.
2. The volume of the euphotic zone.
3. The extent of biological activity.
4. The detention time within the basin.
5. The level of the penstock in the reservoir or level of discharge from a lake.

Preparatory to the investigation of a lake or reservoir, an accurate map of the drainage basin should be obtained. Such a map will give an indication of the area of land drained and thus the land contribution to the lake water quality. Potential sources of pollution should be indicated on such a map.

Whenever possible a contour map of the lake or reservoir basin will be of tremendous advantage to the investigator. A contour map aids in determining sampling location, and it also provides for the calculation of the volume of water within a particular depth stratum. This is important in ascertaining the volume of water that may be representative of a particular set of conditions found through sampling. For example, abundance in aquatic productivity is found in the littoral area in association with the euphotic zone.

Additional physical information that will serve as essential background material for a lake investigation includes the following: area, mean depth, maximum depth, area of various depth zones, volume of depth strata, shore length, shore development, littoral slope, number of islands, area of islands, shore length of the islands, drainage area, rate of runoff, average inflow, average outflow, detention time, water level, and water level fluctuations.

Sampling locations on a lake or reservoir should be located near the mouth of all major inflowing streams, as well as the lake or reservoir outfall. The effluent from a natural lake will provide information on the water quality of its epilimnion. Within the lake or reservoir, a number of sampling locations may be chosen depending upon the problem under investigation and the conditions to be studied. An investigation of the kinds and relative abundance of aquatic vegetation would be limited to the littoral area. A mapping of aquatic plants with an indication of predominant species and their relative abundance often proves useful for future comparison to record relative changes in the vascular plant population. A mapping of the deep water attached plants such as *Chara* may be accomplished with the use of one of the sampling dredges. Mapping the vegetation generally can be accomplished satisfactorily with a boat reconnaissance survey, and frequent inspection stops, in the littoral area during the peak of the summer vegetative growing season.

Fish sampling is more profitable in shallow water areas. Such sampling can be accomplished with the use of electro-fishing devices, seines, hoop nets, or gill nets set in the region of the upper thermocline. The latter will sample a fish population not usually observed in shallow water areas.

The use of transections in sampling a lake bottom is of particular value because there are changes in depth and because benthos concentration zones usually occur. Unless sampling is done systematically, concentration zones of benthic organisms may be missed entirely or represented inadequately. Maximal benthic production may occur in the profundal region. Because depth is an important factor in the distribution of bottom organisms, productivity is often compared on the basis of samples collected from similar depth zones.

A circular lake basin should be sampled from several transections extending from shore to the deepest point in the basin. A long narrow basin is suitable for regularly spaced parallel transects that cross the basin perpendicular to the shore beginning near the inlet and ending near the outlet. A large bay should be bisected by transections originating near shore and extending into the lake proper.

Transections aid also in sampling the plankton population. Because of the number of analyses necessary to appraise the plankton population, however, more strategic points are usually sampled such as water intakes, a site near the dam in the forebay area or discharge, constrictions within the water body, and major bays that may influence the main basin. Because of significant population variations, plankton samples must be taken vertically at periodic depths and at different times over the 24 h day. Vertical samples for chemistry analysis and for plankton should be taken from the deepest portion of the lake or reservoir basin as a minimum. As a minimum also, two sampling locations should be in the epilimnion, at least one in the thermocline, and an additional two locations to represent the hypolimnetic waters.

There are definite advantages in sampling the benthic population in winter beneath ice cover when samples can be collected at definite spaced intervals on a transection and the exact location of sampling points can be determined. Collections at this time are also at the peak of the benthic population when emergence of adult insects does not occur.

Reservoirs usually are long and narrow water bodies with the widest portions occurring downstream near the dam. They are particularly suited for the placement of imaginary transection lines that extend perpendicularly from one shore to the opposite shore. Sampling locations can be conveniently sited on these transections.

The frequency of sampling a lake or reservoir environment depends upon the objectives of the study. When possible, samples should be collected during each climatic season to depict seasonal changes in chemical substances, temperature, dissolved oxygen, nutrients, and biological populations. Sampling during the spring and fall overturns provides valuable information, because there is complete mixing of the water body.

14.9.5 STREAMS

Stream sampling should occur in locations where wastewaters are well mixed with stream waters. Seldom is it necessary to sample at various depths in a stream because of incomplete vertical mixing. Sampling usually is either at a 5 ft depth or mid-depth, whichever is less except for certain biological samples such as plankton that may be sampled at a depth of about 1 ft or bacteria that usually is sampled just beneath the surface. Lateral mixing on a stream poses yet another problem and it is good practice to collect samples at quarter points across the stream unless a predetermination of mixing has shown that a single sample at midpoint of the main current is adequate.

Samples should be obtained from major tributaries. As a general rule, tributaries with flows greater than 10 to 20 percent of that of the main stream should be sampled. Tributaries with smaller flows should be included if they are suspected of being a significant contributor of a characteristic that will influence stream quality.

Sampling locations should be upstream and downstream from suspected pollution sources, from major tributary streams, and at appropriate intervals throughout the stream reach under investigation. The upstream or control location should depict conditions unaffected by a pollution source or tributary. The nearest sampling location downstream from the pollution source should be such that it leaves no doubt that conditions depicted by the sample can be related to the cause of any environmental change. The minimal number of locations from this point should be in the most severe area of the zone of active decomposition, downstream in an area depicting less severe conditions within this zone, near the upper upstream reach of the zone of recovery, near the downstream reach of the recovery zone, and in the downstream reach that first shows no effect from the suspected pollution source. Precise sampling locations will depend on the flow, the strength, volume, and types of pollution suspected of entering the watercourse.

15 The Environmental Career

15.1   KEYS TO CAREER SUCCESS

The keys to career success are the same in government service, industry, and consulting activities. They are:

- Diligent work habits.
- Communication ability.
- Knowledge of job-related subject matter.
- Personal drive to present and publish information.

15.2   DILIGENT WORK HABITS

It goes without saying that diligent work habits are required for program success and for career enhancement. Diligent work habits include putting in a full day of work plus some extra time. It means getting to work on or before the office starting time. And it means arriving on time for scheduled meetings and other appointments. Diligent work habits are always noticed by the supervisor.

15.3   COMMUNICATION ABILITY

Communication ability may need to be developed, but it is well worth the effort. It entails both oral and written communication. It requires planning, organization, and practice.

Oral communication is the presentation of briefings, or other information to a group. It probably will entail the development of slides or other information aids to focus the group's attention on the principal points of a discussion. Slides or overhead projections should be carefully planned. They must be able to be easily read by the intended audience. They must be able to be understood by the average reader. They should not contain any extra words. The words should convey the concise thought that is intended. The series of slides or overhead projections should guide the reader through the information that the presentation is designed to convey. The attention span of the audience will be limited to no more than 30 min, thus, it is prudent to limit the number of slides or projections and not overburden the attention of the audience.

The presenter should speak clearly in a firm voice to ensure that all in the room can hear without difficulty. Practice the delivery to ensure that the presentation will fit within the planned time schedule. Good grammar is essential. The use of slang expressions is prohibited. A good presentation will hold the interest of the audience regardless of the subject matter.

15.4   INVESTIGATIVE REPORTING

Written communication is a daily occurrence and requirement in government, industry, or consulting. Reports are prepared, letters written, or issue papers developed. Good report writing is a systematic recording of organized thought. It should represent an orderly presentation of the results of an investigation. It should be clear and concise. It may include background or history, data, data interpretation and discussion, conclusions, and recommendations.

When study objectives are crystallized, before the actual study commences, thoughts about the organization of the report should begin. The embryonic mental concept of the report should grow in organization and content as the investigation progresses, each phase of the study satisfying a part of an objective and fulfilling a part of the report.

The report is the end result of all the efforts expended on the study and should signal the beginning of any indicated remedial action. The report should be planned as carefully as the investigation. The report's style will depend upon its intended purpose and the audience to which it is directed. It may be a record of the findings only. It may be an exposition of existing causes and effects and projections to other considerations that reasonably may occur. It may be a prediction of conditions to come and recommendations for actions to be taken. The report should be considered a document that records all essential facts in a study that will help meet the needs of those concerned, such as technical agencies and representatives of various vested interests who may be for or against the conclusions and recommendations.

Too often the message that an investigator could bring to defining a problem is lost in poor reporting. Basic facts become mired in technical explanation. Concise interpretative reporting supported by uncluttered pertinent graphic and tabular materials will offer the reader a sound comprehension of the findings of fact. Presenting information that is understandable and meaningful to scientists, engineers, administrators, and to the general public is a challenge to the investigator in reporting the results of an investigation.

15.4.1   OUTLINE

The first step in developing a report is to make an outline. The outline should be considered carefully and should include all the necessary items in logical continuity. An outline is a structural skeleton on which a good report may be build. It must have a beginning, a middle, and an end. A good outline will avoid any omissions of necessary materials and save much time in report writing.

15.4.2   ORGANIZATION

The title, author or authors, and table of contents pose initial decisions. Long titles should be avoided, but the title must identify clearly the work accomplished. Acknowledgments of aid and assistance usually are placed near the front of the report following the table of contents. A good report reviewer spends a great deal of time on the critique of the report and such efforts should be acknowledged.

The report's introduction should describe briefly the problem and its location, the study objectives, the inclusive dates of the investigation, the authority for the study and by whom the study was performed. It may relate briefly the methods used to conduct the study, but generally such description should be placed in the appendix, particularly when it is lengthy and includes those methods that are not understood as standard procedures within the profession.

The summary of the report should briefly and concisely relate how the study was accomplished and what was found in the investigation. The entire summary should be as brief as possible and yet contain the essentials of the findings of fact. Stringent review and editing always should be employed. The summary should contain those particular facts that will be used to formulate conclusions, and upon which any recommendations will be made.

The report's conclusions should be concise, positive, lucid statements that can be made from an evaluation of summarized data and other observations. They should enhance an understanding of the problem. The report's summary can be a basis for the formulation of the conclusions. The report's narrative and the data it contains must support the conclusions presented.

Just as many tools may be used to conduct an investigation, words are the tools used to convey thoughts in a report. Great care should be exercised to choose the right words in formulating the conclusions and recommendations. Often the summary, conclusions, and recommendations may be the only parts of the report that are read by many of the report's audience.

There is often a difference of opinion among report writers regarding the numbering of thoughts or paragraphs within the conclusions. From the standpoint of conciseness and adherence to a particular thought it may be helpful to number conclusions in consecutive order at least initially. After these have been edited and reedited, the numbers may be removed without harm to the text material.

The report's recommendations contain the words to stimulate the remedial actions to correct the problem that was the cause of instigating the study. These should be developed with great care and sound logic, and preferably numbered in consecutive order. Just as the individual conclusions are supported by statements within the report's summary, each recommendation should be the result of a particular conclusion and each conclusion requiring remedial action should have its correlative recommendation. The summary, conclusions, and recommendations of a report are the building blocks for the presentation of a logical sequence of thought, the purpose of which is to solve a particular problem.

15.4.3 REPORT DEVELOPMENT

A report consists of a number of logical interrelated processes that begin with the planning of the investigation and end with the distribution of the report. It is in the investigation planning phase that the report takes embryonic shape.

The organization of data within the report demands ingenuity and imagination. Data must be arranged in tables. Lengthy detailed tables should be placed in the report's appendix. When placed in a narrative they detract from the reading coherency. Easy-to-follow summary tables, prepared as a digest of the tabular data in the appendix, are helpful in the narrative to explain and substantiate discussions and conclusions. Graphs should be uncluttered and easy to follow. Broad lines to illustrate trends are preferred. As an artist develops a painting with broad brush and broad strokes, the report's author should arrange graphs to portray essential information and should use them sparingly only to underscore principal thoughts.

In developing a report, the who, what, why, when, where, and how questions must be answered at every opportunity. Nearly every paragraph requires an answer to these questions to result in a clear understanding of events.

Definite assertions should be made and non-committal language avoided. The specific phrase should be chosen rather than the general phrase, the definite over the vague, the concrete over the abstract. Qualifying words should be avoided whenever possible. Within a water quality report, e.g., the biological, microbiological, chemical, and engineering reports are segments of the whole; each should complement the other. The successful approach skillfully blends the findings of all disciplines into a cohesive, inclusive, and comprehensive discussion of environmental conditions found and the measures necessary to enhance those conditions to a level acceptable for society's use.

Reading the manuscript aloud is a necessary adjunct to good report preparation. It tends to accent the flaws in rhythm, as well as illogical approaches and conclusions. Often laying the manuscript aside for a few days will give the writer a new insight and a more objective approach. It is no sign of weakness or defeat when a manuscript is in need of major revision. This is a common occurrence in all writing.

15.4.4 REVIEW AND FINAL REPORT

When a report is submitted and is accepted by a reviewer, both the writer and the reviewer assume certain specific obligations. The writer should submit the report for review only after completing revisions and personal editing, and after striving for the best in manuscript preparation. Further, the reviewer should be informed of the report's purpose and the expected audience, if these facts have not been made clear in the report itself. Preferably, a minimum of two manuscript copies should be sent to each reviewer so that the reviewer may retain one copy for files and return the other with comments in the report's margin if that is applicable.

A proper review, on the other hand, entails consideration of the technical message in a report, as well as the manner in which the message is presented. A review can be directed toward editorial comments only, but rarely can a technical review disregard the editorial aspects. Good grammar and technical competency usually are inseparable. A technical reviewer reads a document for clarity, technical accuracy, and to determine whether a dual meaning is present in the sentence or paragraph. The reviewer is obligated to consider the purpose that the report is designed to fulfill; to be constructive, thorough, and helpful with comments; to be certain of accuracy in suggesting changes; to base comments on the technical level and interest of the report's intended audience; to avoid sarcasm, argument, or destruction of the writer's style for the sake of expression in the reviewer's words; and to appreciate always that the purpose of the review is to help the report's writer produce a better report.

Following consideration of all the review comments, the report is ready for final assembly. Graphs, pictures, and summary tables should be rechecked for consistency and subheads and to make sure that they follow the narrative reference to them. The appendix is the place for the detailed procedures and tables. These, too, should be consistent in format, understandability, and sequential arrangement.

Finally, an appropriate cover should be designed for the report. The report cover serves as a wrapping for the package and may influence the recipient's opinion of the report or even prompt a desire to examine the report's contents in more detail. An imaginative cover design will pay dividends. Usually the most effective designs are simple, direct, and representative of the report's subject matter.

15.5   KNOWLEDGE OF JOB-RELATED SUBJECT MATTER

It is not the purpose of this book to delve into this subject. It is assumed that anyone hired for a particular position will have a working knowledge of the subject matter associated with the position.

15.6   PERSONAL DRIVE TO PRESENT AND PUBLISH INFORMATION

The "publish or perish" mentality is seldom associated with government service, industry, or consulting activities. Despite this, however, the idea will be a plus for anyone eager to advance in an environmental career. Presenting information before a group can be a challenge the first time it is done, but it gets easier each time the opportunity presents itself. Consider the merits of making presentations before a local, national, or international audience:

- It teaches self-discipline in the use of words so that street words such as "you know what I mean," "cool," and "you know" can be strenuously avoided.
- It teaches clock discipline to complete the discussion in the allotted time without a last-minute rush.
- It establishes speaker confidence and stature.
- It establishes a reputation within the employee's company and among peers of other companies.
- It enhances name recognition of the presenter, as well as the company represented; the managers of the company know that a good reputation and name recognition are valuable company assets.

Making a successful presentation requires preparatory effort. Slides or overhead projections must be prepared to serve as focal points for the presentation. Slides and overhead projections must be thoughtfully planned. They should not contain words that are not needed to convey the intended thought. They should be crafted to ensure that anyone in the audience can read them with ease. They must be designed to provide essential information to support the presentation. There should only be enough to support the presentation and no more than required to fill the allotted time for the presentation. The attention of the audience can not be expected to focus on the presentation for more than 30 min. Generally, in a presentation before a group professional meeting, the allotted time for a presentation is 15 min with an additional 5 min for questions from the audience.

The presenter should speak clearly and distinctly to ensure that no one in the audience has to strain to hear and understand the spoken word. Practice is strongly recommended to ensure that the presentation is well rehearsed before the presentation.

The recommendation for those dedicated to career success is to volunteer to conduct briefings and presentations as each opportunity is available. Career success will profit thereby.

Opportunities arise also to present articles, results of investigations, or other information for publication. These opportunities should be grasped whenever there is any information that is worthwhile to distribute in written form. Papers that are published by technical journals are always a boost to career success. The attributes of published materials include:

- Name recognition and a reputation for well-written material.
- Recognition by peers as an expert in the field.
- Invitations to present material at regional or national technical meetings.
- Enhanced reputation in the employee's company.

The recommendation is to publish at every opportunity. A successful career will be enhanced as a result and the employee will be rewarded by the experience.

15.7   CAREER OPPORTUNITIES

15.7.1   GOVERNMENT SERVICE

Government service areas include local, state, and federal governments. There are career opportunities in all. The work is enjoyable. It is program oriented. Generally, a recently hired employee is assigned to work in a program area and the employee's energy is devoted to developing and implementing that program. As career advancements come, the employee may be in charge of a program activity and the responsibilities shift to guiding the program through the bureaucratic structure of the organization. Communication in the form of program presentations and writings become very important and may lead to the success or failure of the program effort.

At the local and state levels, programs interface directly with persons interested in or impacted by the program activities. At the federal level, programs are defined in larger geographical areas and the interface is more with local and state jurisdictions. Those jurisdictions, however, represent their local constituencies. In essence, then, any governmental program has its roots in people impacted by it, and its success depends on information dissemination and the support received from the local level.

The complexity of the bureaucracy increases from local through state to federal organizations. Checks and balances are a part of that bureaucracy. Learning to work within the organization may be a challenge at times, but is a necessary part of program development. Coordination of activities with peers is a necessity, and sometimes the degree of program success is a measure of the coordination efforts that have gone forth. Presentations and the development of sound writing skills contribute to successful coordination. All of these lead to the molding of a successful career.

15.7.2 INDUSTRY

The principal concerns of industry relate to permit compliance including sample collection and laboratory analyses, filing of discharge monitoring reports, hazardous waste activities, air pollution, ground water monitoring, dredge spoil disposal, and drinking water supply concerns. Larger industrial corporations generally employ personnel to look after their environmental interests and concerns. Smaller industrial organizations may employ consultants on a part-time or task basis to fulfill this need. Industry may be impacted by environmental regulations that are proposed or made final, and industry may need someone to keep abreast of the *Federal Register* to summarize and evaluate contents that may be applicable to the particular kind of industry being served. Certain industries may have a need to examine receiving water upstream and downstream from an industrial outfall to evaluate water quality based upon chemical water quality or biological indicators of water impairment.

There is an opportunity for environmental career satisfaction in the industrial arena. The range of environmental needs is great, especially in the larger facilities. Many industrial environmental problems have been solved since the attack on industrial pollution began in the 1940s, but other associated environmental problems remain ripe for present day solution.

15.7.3 CONSULTING ACTIVITIES

Survival in the consulting arena requires a constant search for new work. Typical consulting firms that have need of environmentally trained personnel have contracts with industry, municipalities, or government institutions. Construction activities often have need for environmental assessments and documentation related to the National Environmental Policy Act, but this need usually is satisfied through the issuance of Requests for Proposals and the ensuing contracts that result from such needs.

The needs of industry are associated with permit compliance to ensure that there are no violations that will result in fines or adverse publicity; permit issuance and renewals to ensure that the company has all permits required by applicable environmental laws; self-audits to ensure that federal, state, and local environmental requirements are being met; sample collection and other monitoring activities required by various environmental permits; and laboratory testing. Occasionally there is need for special studies and evaluations related to water supply issues, hazardous wastes, or discharges with the potential to affect surface or ground water quality. Consultants may be employed in any or all of these activities, especially by the smaller industries. Larger industrial corporations may employ on-staff environmental personnel to manage these activities.

The needs of municipalities often are similar to those of industry. In addition, there may be waste water treatment plants to oversee, solid waste collection and management, and water supply sources and treatment facilities.

The needs of government are very broad. Service contracts and task-oriented contracts are issued. In service contracts, the contractor often serves as an extension of the office served. As a result, such contracts include the development of technical reports and issue papers; investigations, providing recommendations, and reporting on day-to-day issues that arise; developing responses to inquiries, and recording minutes of numerous meetings. With a service contract, the contractor stands ready to assist with any day-to-day request from the office that is served. In a task-oriented contract, the contractor is provided a task to complete. The task could be a long-term assignment that includes investigation, literature searching, discussion, conclusion development, recommendations, and report development.

If there is thought of starting an individual consulting firm, success requires a reasonable capability to build a sustaining client base from which to operate. This requirement may be alleviated in part if the person starting the firm is a recognized expert or specialist, and the speciality is needed in several environmental endeavors. For example, a specialist in archaeology may be in demand by firms employed to develop environmental assessments, because the firm with the contract may not have such an individual with this speciality on staff.

15.8   HANDY TROUBLE NUMBERS

Air Risk Information Support Center Hotline 919-541-0888
Army Environmental Information Response Line 800-872-3845
Chemical Emergency Preparedness Program Hotline 800-535-0202
Chemical Transportation Emergency Center 800-424-9300
EPCRA Hotline 800-535-0202
FEMA Flood Insurance Maps 800-358-9616
FIFRA/General Pesticide Information 800-858-7378
Hazardous Technical Information Service 800-848-4847
Indoor Air Quality Information Clearinghouse 800-438-4318
National Response Center 800-424-8802
National Institute of Occupational Safety and Health 800-356-4674
National Air Toxics Information Clearinghouse 919-541-0850
NIOSH Technical Information 800-356-4674
NCI Cancer Information Service 800-422-6237
Pollution Prevention Information Clearinghouse 202-260-1023
Radon Hotline 800-767-7230
RCRA/CERCLA/UST Hotline 800-424-9346
Safe Drinking Water Act Hotline 800-426-4791
Solid Waste Assistance Program 800-677-9424
Stratospheric Ozone Hotline 800-296-1996
Substance Identification 800-554-1404
TSCA Asbestos Hotline 202-554-1404
Wetlands 800-832-7828

References Cited and Selected Reading

Bregman, J. I. and Mackenthun, K. M., *Environmental Impact Statements,* Lewis Publishers, Chelsea, MI, 1992.

Ellis, M. M., Detection and measurement of stream pollution, *Bul. U.S. Bur. Fish.,* 48, 365, 1937.

Environmental quality, 25th Anniversary Report, The Council on Environmental Quality, Washington, D.C., 1997.

Generic protocol for conducting environmental audits of federal facilities, Ref. EPA 300-8-96-0123, U.S. Environmental Protection Agency, Washington, D.C., 1996.

Leopold, A., *A Sand County Almanac,* Oxford University Press, New York, 1949.

Mackenthun, K. M., The phosphorus problem, *J. Am. Water Works Assoc.,* 60 (9), 1047, 1968.

Mackenthun, K. M., The practice of water pollution biology, U.S. Department of the Interior, Federal Water Pollution Control Administration, Washington, D.C., 1969.

Mackenthun, K. M. and Bregman, J. I., *Environmental Regulations Handbook,* Lewis Publishers, Chelsea, MI, 1992.

McKee, J. E. and Wolf, H. W., *Water Quality Criteria,* Water Quality Control Board, Sacramento, CA, 1963.

Navy Environmental Compliance Sampling and Field Testing Procedures Manual, NAVSEA T0300-AZ-PRO-010, Department of the Navy, Washington, D.C., 1997.

Pocket Guide to Chemical Hazards, Public Health Service, Center for Disease Control, National Institute for Occupational Safety and Health, U.S. Department of Health and Human Services, Atlanta, GA, 1997.

Selected Environmental Law Statutes, West Publishing, St. Paul, MN, 1996.

Seventh annual report on carcinogens, National Institute of Environmental Health Sciences, Research Triangle Park, NC, 1994.

Shelford, V. E., An experimental study of the effects of gas wastes upon fishes, with especial reference to stream pollution, *Bul. Ill. State Lab. Nat. Hist.,* Champaign, IL, 11, 381, 1917.

The quality of our nation's water, Rep. EPA 841-S-94-002, U.S. Environmental Protection Agency, Washington, D.C., 1994.

Water Quality Criteria, State Water Pollution Control Board, Sacramento, CA, 1952.

Water Quality Criteria, National Academy of Sciences, National Academy of Engineering, Washington, D.C., 1974.

Index

A

Acronyms, 5
Acid rain, 43
Act to Prevent Pollution from Ships, 31, 34
Actinolite, 41
Advisory Council of Historic Places, 71
Air, 3
  Clean Air Act requirements, 37
  hazardous air pollutants, 40
  pollutants, 39
  quality, 3
  regulations, 12
Air Quality Control Regions, 39
Algae, 79
Amosite, 41
Anthophyllite, 42
Applicable or relevant and appropriate
          requirements (ARARs), 56
Asbestos, 39
  control, 42–43
  inspectors, 42
  regulations, 14, 42–43
  sampling, 43
Asbestos Hazard Emergency Response
          Act, 42
Asbestosis, 42
Attainment, 39, 44
Audits, 6, 73
Audit protocol, 75
  questions, 73

B

Bald Eagle Protection Act, 71
Ballast water, 29
  exchange, 29
Best Available Control Technology, 39
Biological toxicity, 26
Blue book, 22
Boron, 105
Boston Harbor, 19

C

Career development, 117
  Present and publish information, 121
  Consulting careers, 123
Carbon monoxide, 3, 39, 44
Carcinogens, 96
Categorical exclusions, 15, 18

Chemical
  hazards, 95
  manufacturing, 59
  substances, 60
Chlorine, 105
Chlorofluorocarbons, 44
Citizens for a Better Environment, 21
Citizen's Suits, 6
Clean Air Act, 37
Clean Air Act Section 309, 16
Clean Water Act, 21–28, 37
Clean Water Action Plan, 28
Cleanup standards, 56
Coastal zone, 71
Coastal Zone Management Act, 71
Cobalt, 105
Code, U.S., 6
Code, U.S. titles, 7
Code of Federal Regulations, 8
Combustion, 44
Communication ability, 117
Comprehensive Environmental Response,
          Compensation, and Liability
          Act, 53
Compliance, 73
Congressional Record, 7
Consulting careers, 123
Control technology, 44
Control Technology Guidelines, 39
Convention for the Prevention of Pollution
          from Ships, 32
Corps of Engineers, 26, 31
Council on Environmental Quality, 3, 16
Crocidolite, 42
Crysotile, 41
Cuyahoga River, 19

D

Density currents, 101
Detroit Lakes, 19
Dissolved oxygen, 102
Downstream drift, 79
Drinking water
  regulations, 12, 30
  sources, 30
  standards, 29
  treatment, 30
Dredge and fill, 26
  regulations, 13
Dredge spoil, 31

E

Ecosystem recovery, 77
Effluent guidelines, 7, 26, 28
Egg deposition, 80
Emergency Planning and Community
        Right-To-Know Act, 61
Emergency Response Plan, 62
Emissions, 45
Endangered species, 14, 64
Endangered Species Act, 64
Enforcement 8, 45, 52
Environment
    clean, 2
    effects abroad, 18
    EPA comments, 16
Environmental
    advantage, 1
    ethic, 4
    investigation, 8
    Reviews, 18
    Study, 18
Environmental Assessment, 15, 18
Environmental Defense Fund, 21
Environmental Impact Statement, 15, 16, 18
    format, 16
    information, 17
Epilimnion, 100
Estuaries, 25
Eutrophication, 79, 104
Executive Order, 9
    11574, 20
    12114, 18
    12856, 61
    12898, 17
Exclusive Economic Zone, 18, 29

F

Farmland Protection Policy Act, 17
Federal Food, Drug, and Cosmetic Act, 69
Federal facilities, 52
Federal Register, 7, 9
Federal Security Agency, 20
Federal Insecticide, Fungicide, and
        Rodenticide Act, 68
Flambeau River, 19
Fish Advisories, 3
Finding of No Significant Impact, 15
Fish and Wildlife Coordination Act, 70
Form R, 62
Fox River, 19
Flow, 103
Friable, 42
Freedom of information, 9

Freedom of Information Act, 18
Fundamental concepts, 5

G

Garbage, 34
Global Commons, 18
Gold Book, 22
Government service careers, 122
Green Book, 22
Great Lakes, 3, 29
Green Bay, 19
Green Bay fly, 19
Groundwater, 3, 29

H

Hazard
    communication, 96
    substances, 24
Hazardous air pollutants, 40, 45
Hazardous waste, 49–53
    audits, 73
    characteristics, 49
    discarded commercial chemical product, 49
    generator, 50
    listed, 49
    manifest, 50
    toxicity characteristics, 49
    transporters, 50
    treatment, storage, and disposal, 51
    regulations, 13
Help hotline numbers, 124

I

Imminent hazard, 60
Incidental taking, 66
Industry careers, 123
Injection wells, 53
International Agency for Research
        on Carcinogens, 96
International Convention for the Prevention
        of Pollution from Ships, 32
Inorganic silt, 106
Invasive species, 29
Investigative reporting, 118
    organization, 118
    outline, 118
    report development, 119
    review and final report, 120
Iron bacteria, 30

J

Job careers, 117

K

Klamath River, 19

L

Lakes, 2
Lake aging, 78
Lake Erie, 19
Lake Sebasticook, 19
Land, 49
Laws, 10, see also specific listings
    Act to Prevent Pollution from Ships, 31, 34
    Asbestos Hazard Emergency Response Act,
        42, 59
    Bald Eagle Protection Act, 71
    Clean Air Act, 16, 37
    Clean Water Act, 21–28
    Coastal Zone Management Act, 71
    Comprehensive Environmental Response,
        Compensation, and Liability Act, 53
    Convention for the Prevention of Pollution
        from Ships, 32
    Emergency Planning and Community
        Right-To-Know Act, 61
    Endangered Species Act, 64
    Farmland Protection Policy Act, 17
    Federal Food, Drug, and Cosmetic Act, 69
    Federal Insecticide, Fungicide, and
        Rodenticide Act, 68
    Federal Water Pollution Control Act
        (1977), 21
    Fish and Wildlife Coordination Act, 70
    Freedom of Information Act, 18
    Hazardous and Solid Waste Amendments of
        1984, 49
    International Convention for the Prevention of
        Pollution from Ships, 32
    Marine Mammal Protection Act, 65
    Marine Protection, Research, and Sanctuaries
        Act, 31
    Migratory Bird Treaty Act, 71
    National Environmental Policy Act, 15
    National Historic Preservation Act, 71
    National Invasive Species Act, 28
    Nonindigenous Aquatic Nuisance Prevention
        & Control Act, 28
    Ocean Dumping Ban Act, 32
    Occupational Safety and Health Act, 95
    Oil Pollution Act (1912), 20
    Oil Pollution Act (1924), 20
    Oil Pollution Act (1980), 24
    Pollution Prevention Act, 61, 81
    Public Health Service Act (1912), 20
    Public Law 660, 20
    Public Law 92–500, 21
    Resource Conservation and Recovery Act, 49
    River and Harbor Act of 1899, 20
    Save Drinking Water Act, 29
    Toxic Substances Control Act, 59
    Water Pollution Control Act (1948), 20
    Water Quality Act, 20
Lead, 3, 39
Legislative history, 9
Leopold, Aldo, 4
Light in water, 108
Local Emergency Planning Committee, 61
Lung cancer, 41

M

Madison lakes, 19
Marine mammals, 65
    regulations, 14
Marine Mammal Commission, 66
Marine Mammal Protection Act, 65
Marine Protection, Research, and
        Sanctuaries Act, 31
MARPOL, 32
    Special areas, 33–34
Material Safety Data Sheets, 45, 63, 96–97
Maximum contaminant levels, 31
Maximum contaminant level goals, 31
Mayfly larvae, 19–20
Medical wastes, 32
Mesothelioma, 42
Methyl isocyanate, 61
Migratory Bird Treaty Act, 71
Mississippi River, 19
Mixing zone, 32
Molybdenum, 105
Monitoring
    National Air Monitoring Stations, 39
    State and Local Air Monitoring Stations, 39

N

National Air Monitoring Station, 39
National Ambient Air Quality Standards, 39
National Contingency Plan, 54
National emission standards, 39
National Emission Standards for
        Hazardous Air Pollutants, 39
National Environmental Policy Act, 1, 3, 12,
        15–16, 43
    Section 309 EPA comments, 43
National Historic Preservation Act, 71
National Institute for Occupational Safety
        and Health, 95
National Invasive Species Act, 28

National Priority List, 53–54
National Register, 71
National Technical Advisory Committee, 22
National Water Quality Inventory, 2
Natural Resources Defense Council, 21
New Source Performance Standards, 39
Nonindigenous Aquatic Nuisance Prevention
        and Control Act, 28
Nitrogen, 28
Nitrogen oxide, 3, 39, 45
Nonattainment, 39, 45
Nonpoint sources, 25
Nonindigenous species, 28
Northern Right Whale, 67
Nutrients, 78, 104

O

Occupational Safety and Health Act, 95
Occupational Safety and Health
        Administration, 95
Ocean dumping, 31
    mixing zones, 32
    permit, 31
Ocean Dumping Ban Act, 32
Offset, 45
Oil Pollution Act, 20
Okeechobee Lake, 19
Organic pollution, 106
Overseas Categorical Exclusion, 18
Overseas Environmental Assessment, 8
Overseas Environmental Impact Statement, 18
Oxygenated fuel, 45
Ozone, 3, 46
Ozone hole, 46

P

Particulate matter, 39, 46
Permissible exposure limit, 42, 96
Permits, 9, 25–26, 28, 46
    air, 46
    discharge, NPDES, 26
    hazardous waste, 51
    ocean dumping, 31
Pesticides, 68–70
    general use, 69
    restricted use, 69
Phosphorus, 28, 104
Photosynthesis, 103
Plastic, 35
Polychlorinated byphenyls (PCBs), 60
    regulations, 14
Pollution prevention, 81
    use minimization, 82

training, 82
    plans, 63, 81, 82
Pollution Prevention Act, 61
Potomac River, 19
Primary Drinking Water Standards, 29
Public laws, 10, see also specific listings
Public law 660, 20
Public law 92–500, 21
Public Health Service, 20
Public Health Service Act, 20
Public vessel, 52

Q

Quality Criteria for Water, 22
Questions & answers, review, 83–88

R

Record of decision, 56
Recovery from pollution, 77
    lakes, 78–79
    streams, 78
Red book, 22
Regulation, 4, 11–12, 26
Regulatory process, 27
Remedial action, 54
Reorganization Plan #2, 20
Reorganization plan #3, 20
Reservoirs, 100
    main stream, 100
    storage, 100–101
Resource Conservation and Recovery Act, 49
Review answers, 88
Review questions, 83
Risk assessment, 14
River and Harbor Act (1899), 20

S

Safe Drinking Water Act, 29
Safety, 95
Sebasticook Lake, 19
Secondary Drinking Water Standards, 29
Sensitive ecological areas, 25
Settlement Agreement of 1976, 21
Sewage, 25, 33
Silicon, 105
Site discovery, 55
Small quantity generators, 50
Soil Conservation Service, 17
Sole Source Aquifers, 29
Solid waste, 3
    regulations, 13
Special areas, 33–34

Standards for chemical carcinogens, 96
State certification, 25
State Emergency Response Commission, 61
State and local air monitoring stations, 39
State implementation plans, 39, 46
Stormwater, 21
Stream pollution zones, 107
   active decomposition, 107
   clean water, 107
   degradation, 107
   recovery, 108
Submarines, 35
Sulfur dioxide, 3, 39, 47
Superfund, 52
   preliminary assessment, 55
   site discovery, 55
   site inspection, 55
   record of decision, 56
   regulation, 13
   remedial investigation/feasibility study, 56

T

Taking issue, 65, 68
Tastes and odors, 30
Temperature inversion, 47
Temperature as a regulator, 99
Testing chemical substances, 59
Thermocline, 100
Thermal stratification, 100
Threatened species, 65
Toxicity characteristics, 50
Toxic
   chemicals, 62, 64
   pollutants, 23
   release inventory data, 62
Toxic Substances Control Act, 59
Train, Russell, 1
Tremolite, 42
Trophogenic zone, 103

U

Ultraviolet B, 47
Underground injection, 29

Underground storage tanks, 53
Uniform national discharge standards, 25
Upstream migration, 79

V

Volatile organic compounds, 47

W

Washington Lake, 19
Water, 19
   Clean Water Act requirements, 37
   regulations, 12
Water Pollution Control Act (1948), 20
Water quality
   criteria, 21
   density, 99
   estuaries, 3
   investigations, 108
   lakes, 2
   rivers, 2
   standards, 21, 28, 31
   state reports, 2
   temperature as a regulator, 99
Water Quality Act of 1965, 20
Water quality investigation, 108
   benthic sampling, 112
   data collection, 112
   information collection, 111
   objectives, 108
   planning, 109
   lake studies, 113
   stream studies, 115
Wells, 29
   injection, 53
Wellhead protection, 29
Wetlands, 3

Z

Zebra mussels, 28
Zinc, 105

Milton Keynes UK
Ingram Content Group UK Ltd.
UKHW040051071024
449327UK00019B/476